BENEFICIAL AND PEST INSECTS

The Good, the Bad, and the Hungry

The Hungry Garden series #3

Rosefiend Cordell

Rosefiend Publishing.

Ordering information: For details, contact the publisher at hello@melindacordell.com
Cover design by Melinda R. Cordell
Book formatting by Melinda R. Cordell

D2D paperback: 978-1-953196-65-1
First Edition: June 2021

10 9 8 7 6 5 4 3 2 1 blast off!

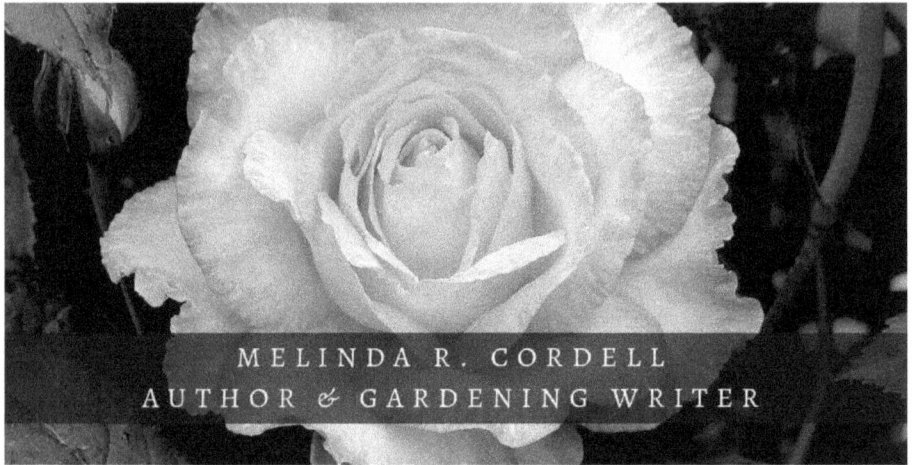

MELINDA R. CORDELL
AUTHOR & GARDENING WRITER

For more information (and books!), visit my website at
https://melindacordell.com/

Subscribe to my Newsletter
and get a free gardening book.

The Hungry Garden Series

Big Yields, Little Pots – Container Gardening for the Creative Gardener
   Book 1

Edible Landscaping – Foodscaping and Permaculture for Urban Gardeners
   Book 2

Beneficial and Pest Insects – The Good, the Bad, and the Hungry
   Book 3

Indoor Gardening – Growing Herbs, Greens, & Vegetables Under Lights
   Book 4

FORTHCOMING BOOKS!
Growing a Food Forest – Trees, Shrubs, & Perennials That'll Feed Ya!
   Book 5

Wildscaping – Using Native Food Plants to Create an Ecologically-Friendly Garden
   Book 6

Survival Rations! – Foraging in Wild Spaces for Greens, Berries, & Nuts
   Book 7

Victory Gardens – We Can Grow It!
   Book 8

Do you have a gardening story you'd like to share? Tell me your experience with bugs in the garden (or any other gardening stories) at hello@melindacordell.com and I'll share it in a future book!

TABLE OF CONTENTS

A Quick Overview of the Insects in This Book
And the Order They Belong To

Odonota
Dragonflies and damselflies

Orthoptera
Grasshoppers
Crickets

Diptera
Fungus gnats
Mosquitoes
Flies

Hemiptera
True bugs
Aphids
Tarnished plant bug
Four-lined plant bug
Big-eyed bugs
Squash bugs
Stink bugs
Scale
Whitefly
Mealybugs
Wheel bugs
Assassin bugs

Thysanoptera
Thrips

Coleoptera
Beetles
Colorado potato beetle
Cucumber beetle
Ladybugs
Mexican bean beetle
Flea beetle
Japanese beetle
Asparagus beetle
Wireworms
Weevils
Emerald ash borer

Hymenoptera
Ants, bees, wasps
Wasps
Cicada killers
Leaf-cutter bees

Lepidoptera
Butterflies and moths
Cabbage maggot
Cutworms
Bagworms
Leaf miner
Codling moth
Corn borer
Corn earworms
Cabbage caterpillars
Tomato hornworm

Neuroptera
Lacewings

Mantodea
Mantids
Praying mantis

Thysanoptera
Thrips

Hemiptera
True bugs
Aphids
Tarnished plant bug
Four-lined plant bug
Squash bugs
Stink bugs
Scale
Whitefly
Mealybugs

CRITTERS THAT ARE NOT INSECTS

Arachnids (includes spiders, mites, scorpions, and most recently, horseshoe crabs, no lie)

Mites
Spider mites

Gastropods
Snails and slugs

INTRODUCTION

Oh, These Are the Insects in Our Neighborhood

I had a lot of favorite classes during my undergraduate years in college (I was an unabashed nerd and loved learning new things), and one class that really stuck with me through the decades was my entomology class, taught by Dr. Johanne Wynne Fairchild.

Really, Dr. Fairchild was a powerhouse. She covered a lot of the key horticulture classes I had in school and also taught in the Biology department. But her entomology class was great fun. We had to collect insects for a collection, naturally. I used a pint-sized Mason jar that I'd found in my Great-grandma Lillian's root cellar that had a cotton ball dipped in rubbing alcohol to kill insects, or I'd put them in a plastic bag and stick them in the freezer. I gathered insects everywhere. I got an earwig in a box of apples at the local apple orchard; I caught a sphinx moth (after class was over, I kept that specimen mounted in an old plastic watch box because it was that pretty).

The taxonomy of insects fascinated me. That was one thing that was definitely a pleasant surprise in my college classes. I knew what taxonomic keys looked like because I used to read *Flora of Missouri* by Julian Steyermark in high school, and had tried to use the taxonomic keys to identify plants but was absolutely bamboozled by them. Those things looked scary, and they still do to me, not gonna lie. But we learned to use them in this class and others, and they're a valuable key when you're out in the field.

And to me, being able to categorize insects, plants, and other creatures, and being able to fit different creatures into specific categories, is something that I've gotten a lot of use out of through the years.

When you're dealing with insect pests, knowing what order they're from helps you target them more effectively. Coleoptera, Diptera, Lepidoptera – the beetle family, the fly family, the butterfly family – each of these insect families have a larval stage in their development. You can target the larval stage in these insects to catch them when they're still vulnerable, thin-skinned, and squishy.

What's more, controls that work for one type of larvae will also work for others. I had problems with fungus gnats in my houseplants. Then I found a very simple solution. Mosquito dunks – those floating disks you put in your garden pond to kill off mosquito larvae – also work against fungus gnat larvae. You have to use the mosquito dunks that contain BT, which is *Bacillus thuringiensis*, a soil bacteria that creates proteins that are toxic to insect larvae.

Basically all I did was put a chunk of mosquito dunk in my watering can. Every time I water my houseplants, they get a dose of BT, which is harmless to the plants, but kills the fungus gnat larvae that live in the soil.

This control works against both mosquito larvae and fungus gnat larvae, because both of these insects are in the fly family (Diptera), and they share the same vulnerabilities against BT.

Look, I am a straight-up naturalist nerd. I'm fascinated by animals and plants and little critters. You don't have to be – though I think it makes life more funner to totally nerd out. I've found it helpful through the years to understand the whys of how the natural world works.

At any rate, here's a book about bugs in YOUR garden. You should totally get to know the creatures that make your garden their home. Some are evil beyond words!! Some are pretty good. Some are in the middle there, have their good points and not-so-good points, but they're all a part of the busy world of life that lives on your roses, cantaloupes, radishes, and in your soil, and makes the world an interesting place.

Let's get started!

Japanese beetles eating up my Penelope roses. Grr.

INTEGRATED PEST MANAGEMENT

A Handy Guidebook to Pest Management

Integrated Pest Management is a ten-dollar name for a way of dealing with pests that just makes a lot of sense.

Integrated pest management, abbreviated as IPM, is basically a mindful approach to pest control. You're not out to wholly eradicate pests, as chemical companies have urged in the past (also, to be honest, pest eradication is a good way to sell chemicals). The aim here is control – keeping the pests at levels low enough where the damage they do is minimal or not even noticeable.

7

Pest *eradication* ends up destroying your insect allies and adversely disrupting the local ecosystem, and adds unnecessary chemicals to the groundwater and soil.

Additionally, because I am a cheap so-and-so, I'd like to point out that those damn chemicals are expensive. Why buy chemicals when I could buy a new rose, or a book, or all those fries that my kids keep demanding every night when I leave work? We have priorities here.

At any rate, IMP at its best is basically trying to figure out the best means of controlling an insect pest, whether through chemical or biological controls, or both. If chemicals are used, they're used responsibly – spray only what you need.

When I was a municipal horticulturist, I used Round-Up to keep up with the weeding, but I was careful. Instead of spraying wide swaths of ground with the chemical, I kept the spray nozzle super-close to the weeds, so the spray went mainly on the weeds, and in such a way that only the barest minimum of spray reached the ground. With small weeds, I would wipe the nozzle on the leaves, leaving a line of poison on the leaves for the plant to absorb. The weeds died (and if they didn't, they would after I stopped by for a second round of spraying).

By using the chemicals in this way, I saved money, and I kept the environmental impact of the chemical down. Round-Up is bad for earthworms, and I really wanted to keep the worms in the garden soil undisturbed as much as possible. Generally I would hoe or pull weeds, and I mulched a lot to keep the weeds down, but if there were too many weeds, or if my to-do list was particularly arduous, I would fall back on the spray.

The six parts of IPM

IPM is broken down into six parts:

1) Acceptable pest levels.

Completely eradicating an insect species from an area is difficult and costly (both in terms of money and environmental impact). With good IPM, you're focused on *control*, not eradication. You're allowing a few pests to show up here and there and being comfortable enough to say "Okay, no big deal."

So you rub some aphids off the stems of your roses, and if there are only a few of them and the rose isn't looking wilty and sad, then you don't worry about them. You know to keep an eye on the aphid population to make sure they stay at low levels. This course of action is better than rolling out a barrel of poison every time you see a couple of aphids.

Though I totally understand about wanting to blast these little jerks with poison.

2) Preventative cultural practices

This is an aspect of IPM that should get a lot more press, but it probably doesn't because it's kind of hard to put into words.

Eliot Coleman wrote about using a "plant positive" approach instead of a "pest negative" approach. That is, the best way to create a contented garden is to focus on giving the plants all the things they need to live their best life.

The best way to attain this for your plants is by adding compost and mulch to the soil to develop a bustling ecosystem around the plant's roots.

Note that building a fertile soil is a process. If you are starting with a mostly dead soil, it will take a little extra time and work to bring this about. If your soil is already fertile, you'll still need to maintain soil fertility through adding more compost and mulch as time goes by.

Once you have a fertile soil and healthy plants, many pest problems will take care of themselves. Soil health is not a cure-all – but it's an excellent place to start.

Other preventative practices include selecting the best plants for your particular climate. For instance, English roses, lupines, and primroses, which grow so well in the United Kingdom, are much more of a challenge in my home state of Missouri. Conversely, the native coneflowers and grasses of the Great Plains find the cool and misty English climate something of a challenge.

Another aspect of preventative practice includes good sanitation – things like quarantining your plants if they become diseased – promptly removing diseased or dying plants – and cleaning your pruning shears as you move from plant to plant in the garden. Many plant professionals use 70% isopropyl alcohol for disinfecting pruners. Have a small, covered container of the alcohol handy, and as you move to another plant, wipe the blades of your pruners with the damp cloth.

When I worked in the rose garden, which was badly afflicted with rose rosette disease, I used a solution of one part bleach to nine parts water to wipe down my Felco blades every time I moved from one rose bush to the other.

One further way to keep disease transmission down is to avoid working with wet plants. Disease can be easily spread through water droplets carried from one plant to another. Spores and bacteria don't float through the air on their own. But if they get picked up by a drop of water, they get trapped in there by the surface tension of the water droplet. If you touch that rose, then your hand picks up those droplets, and then those droplets end up on the next rose you touch. Boom, disease transmission.

Look at those delicious and yummy DISEASE VECTORS!!

3) Monitoring/Observation

The third tenant of IPM is observation, which is done through a number of methods.

During inspections, look for tiny webs on or under the leaves (spider mites), little critters that look like mushy cotton (mealybugs), little pear-shaped bugs in green or yellow (aphids), specks of white that fly (whiteflies), little round scabs that you can scrape off (scale), or gnats that live in the soil that like to fly directly at your eyes (fungus gnats).

Look under the leaves and around the leaf stems for any unwelcome guests that like to hide out in these areas.

You can set up insect traps to keep an eye on the different insects that are visiting your garden; you can look closely at your plants as you're working with them.

If you keep a gardening notebook (I do hope you are keeping a gardening notebook), keep a section in it dedicated to when you see specific insect pests or diseases. At the beginning of the year, mark these dates in your planner, so then, when those dates roll around, you can keep an eye out, and stop the diseases and pests before they get a foothold.

It's a lot easier to catch and control a potential infestation if you nip it in the bud.

Bagworms are a great example. If you spray for bagworms when they're still small, they're easier to kill off because they haven't had a chance to spin a full-sized protective cocoon. When they're nearly full-grown, they can hide inside their waterproof cocoon bag and pull it tightly shut every time you try to spray them. But small bagworms have a small cocoon that their heads stick out of, so you can spray the heck out of them.

Baby bagworm. The bag they live in when they're small is not big enough to fully close yet, so it's easier to get insecticidal spray on the little jerks.

4) Mechanical control

Insects can develop immunity to all kinds of sprays and poison in the garden. But do you know what they can't develop immunity to? Your fingers squishing them or dropping them into a can of soapy water.

It's a good idea to occasionally wash your houseplants with warm, soapy water to get rid of dust, insects, honeydew from the insects (honeydew is sugar water that insect pests secrete on the leaves) and old leaves. Rinse with plain water, then let the plants dry. Don't wash African violets or gloxinias; they resent this treatment – they're like cats when it comes to water.

Mechanical controls include squishing, hand-picking, traps, tillage, and vacuuming. If pests are really getting you down, try sucking them up with your leaf blower set to reverse. Rubber-band a piece of nylon over the opening so they don't clog your machine from the inside. Then suck up a bunch of pests, shut off the machine, then knock them off the nylon into a bucket of soapy water. Repeat. This works great for potato beetles.

5) Biological control

This includes encouraging beneficial insects to your garden. Add plants they like into the garden, hang up a pollinator house for solitary-nesting native bees, and create small habitats for helpful insects to live in. You could also use biological insecticides, or buy predatory insects (which sometimes simply fly away – but they're helpful somewhere).

It also helps to know which insects to kill and which ones to encourage – which is what I hope to do in this book.

6) Responsible use: When using pesticides, remember these important words:

Always! Read! And! Follow! Label! Directions!

When using a chemical agent, use only as much as you need, and make sure it is for the critter you are targeting – that is, don't use something formulated to kill slugs to try and kill beetles.

7) One additional note

A great way to control insect outbreaks is to understand the life cycle of the insect you're trying to stop. By looking at their life cycle, you can find ways to disrupt the insect's breeding or growth. So you can look at the flea beetle's life cycle and think, "So if the larvae are in the soil right now, I can drench the surface of the soil with soapy water every week and kill them off while they're still underground." So you upend a bucket of soapy water on the plant every week, making sure you get the leaves while also soaking the ground around the plant. That's how you make it a two-fer.

Time and again, I've been able to use this understanding to stop insect outbreaks in their tracks, though it helps that I took Dr. Johanna Fairchild's Entomology class in college to show me how. *tips hat*

Quick Tips About Insecticides

Most of these insects can be controlled by keeping the plant clean through an occasional bath. However, if an infestation starts, get a container of insecticidal soap and add just a little rubbing alcohol to it. The alcohol penetrates the bug's protective waxy coating, allowing the soap to kill it. Spray this mixture weekly on your plants until the infestation is gone.

However, there are several plants, such as impatiens, begonias, jade plant and fuchsias that will not react well to insecticidal soap. If in doubt, spray just a few leaves with the insecticidal soap and check the leaves after a day or two to see if they're affected by it.

If you must use an insecticide, be sure that it's labeled for the specific insects that you're targeting, and that you always read and follow label directions!

Other killing sprays you can use include pyrethrin and insecticidal soap. These sprays, however, kill on contact only; they are not residual. It is satisfying, though, to shake the plant, and when the whiteflies swarm up, spray a cloud of pyrethrin and watch them sink back down.

You could also use a systemic insecticide. This is a granular poison you dig into the soil of your plant. When you water, the plant absorbs the poison, and when bugs drink the plant juices, they get poisoned. However, if you have pets that eat leaves, or little kids, do not use systemics.

Kaolin Clay for Pest Control

One effective control against stink bugs, flea beetles and other leaf-eating bugs is kaolin clay.

Kaolin is a type of clay often used in kitty litter (if I'm remembering my Soils class correctly). But when kaolin clay is finely ground and combined with water in a 1 to 6 percent concentration, it's been proven to act as an effective barrier to pests, and also protects the leaves from sunscorch on hot summer days. The reflective white film that kaolin forms over the leaves actually helps photosynthesis while reducing heat stress. Kaolin clay is used in orchard production, and can be effective in the home garden to repel insects such as stink bugs and many other plant-eating insects. Be sure to completely cover both surfaces of the leaf when spraying this, both upper and lower.

Now in one study, when kaolin clay was used on cotton plants, an aphid infestation actually got worse. So as with any preventative measure, it's not perfect. But this sounds like a really helpful option. Try it out with your other pest-control measures, and see what kinds of results you get.

Birds Are on Your Side

One day I was surprised to see a lot of sparrows in the rose garden, hopping around the canes inside the roses. This is odd behavior for house sparrows – usually sparrows will congregate in a large bush that provides cover, or they'll hop around in the open. So I stopped to watch a female sparrow hopping from cane to cane inside a rose bush. With a dart of her head, the female sparrow grabbed something off a leaf and flung it to the ground, then hopped down after it. It was a small green caterpillar, which she quickly ate up. Other sparrows were doing this too.

So the sparrows had noticed that my roses had a caterpillar outbreak and were cleaning them up! I was amazed and grateful. Thanks to the sparrows, there were no signs of caterpillars after that. I was glad I hadn't sprayed for insects.

Somebody should have told me how addicting bird watching is. I bought a copy of Birds of Missouri, which is available from the Missouri Department of Conservation, because I was outside a lot and simply wanted to know more about birds.

One afternoon I was shutting off the water in the rose garden when a tiny slate-blue bird perched on the 'Las Vegas' rose about eight feet away. It sat there for about a minute, observing me. I wrote down its description and found out later that it was a Northern Parula, which is a kind of warbler. I'd never heard of it.

Now I'm scanning the skies and listening to my bird calls tape in hopes that I can identify more species. I have check marks in the Birds of Missouri book to keep tracks of which birds I have positively identified. I'm going with the flow. I learn a lot when I

let myself get excited about a new subject and follow through.
Birds are fun.

A bluetit with caterpillar – not a parula. We don't have bluetits where I live, alas.

Spot-Spraying Insects is Best

I don't even spray insects. I just don't! I gave up spraying insecticide after a tragic (it wasn't tragic for me) experience.

When I first started at the rose garden, the previous horticulturist advised me to spray the roses with insecticide every week. So I made up a full sprayer of insecticide and got to work. As soon as I started spraying a ladybug popped out from behind a rose leaf and died. "Sorry!" I said, too late.

Then a praying mantis fell out of a rosebush with a death throttle, dropping the insect she was eating. "Oh no!" I said.

A flock of lacewings, also good guys in the fight against bad insects, shimmered up and perished in the spray, and a honeybee tumbled out of a rose blossom.

"I'm killing all the good bugs!" I wailed. I then returned the sprayer to the tool shed and swore off spraying insecticide. Oh well, that was one less task on my never-ending list o' things to do. Let the good insects do all the work. Fine with me!

So I stopped buying bulk insecticide. Instead I bought a spray bottle of pyrethrum, or one of insecticidal soap (I'd generally grab one or the other when I stopped by the local nursery). When I saw aphids on my roses, I'd spot-spray the bugs that needed spraying. Generally an application or two would do the trick. I also would rub them off with my fingers if I didn't have a bottle handy. This approach saved money, it saved time, and it saved the beneficial insects.

Now in the case of Japanese beetles, it's worth making a gallon of BioNeem spray and knocking those damn bugs out of the ranks of the living. I hate the way that they will mob a rose blossom.

If you have chickens, Japanese beetles are a great feed supplement. My hens love them. I hang a pheromone Japanese beetle trap in the chicken pen, then, every day, empty the bag of beetles into a shallow pan of water to keep the beetles from flying off, and let the chickens eat them up. My hens come running when I bring them Japanese beetles, and after a couple of quick snaps with their beaks, the beetles are history. If you have more beetles than the chickens can eat, freeze them – they'll keep!

More about Japanese beetles later in this book.

It's a Bug-Eat-Bug World

The beneficial insects page in a horticultural supply catalog is not for sissies. These pages, which list predatory insects that the organic gardener can use to control pests, can make the gardener feel like she's fallen into a Stephen King novel. These beneficial insects are vicious, devious … and boy, are they hungry.

For instance, protozoans called *Nosema locustae* are used to control 90 percent of grasshopper and cricket pests. The protozoa are placed on bran flakes, which unsuspecting young hoppers eat. Once the protozoa infects them, other grasshoppers will seek out these infected hoppers and cannibalize them. Then *they* get cannibalized, and so on. It's like a grasshopper Walking Dead.

Or perhaps you'd prefer some fly parasites, also known as trichogramma wasps. The female wasp lays her eggs inside the pest fly pupae. When the eggs hatch, the tiny wasp larvae consume the fly pupae from the inside.

I apologize to those of you who, upon reading that, spit out your coffee.

Cicada killer wasps, and other wasps, do the same thing. Cicada killers, which are huge wasps (but they aren't mean and are actually kind of fun to play with) will dig burrows going three to five feet into the ground. At the ends she builds a little chamber. Then she goes out, catches some cicadas, stings them to immobilize them, and drags them, still alive, into the chamber. She'll lay an egg on them and seal up the chamber – and when the wasp egg hatches, the larvae will have plenty of food to devour.

The insect world is cruel and vicious, which gardeners can use to their advantage. Organic means of pest control allows gardeners to control pests safely and effectively, without having to use harsh chemicals that kill or adversely affect the living things in the area. If you have an insect pest, it's best to seek an organic solution.

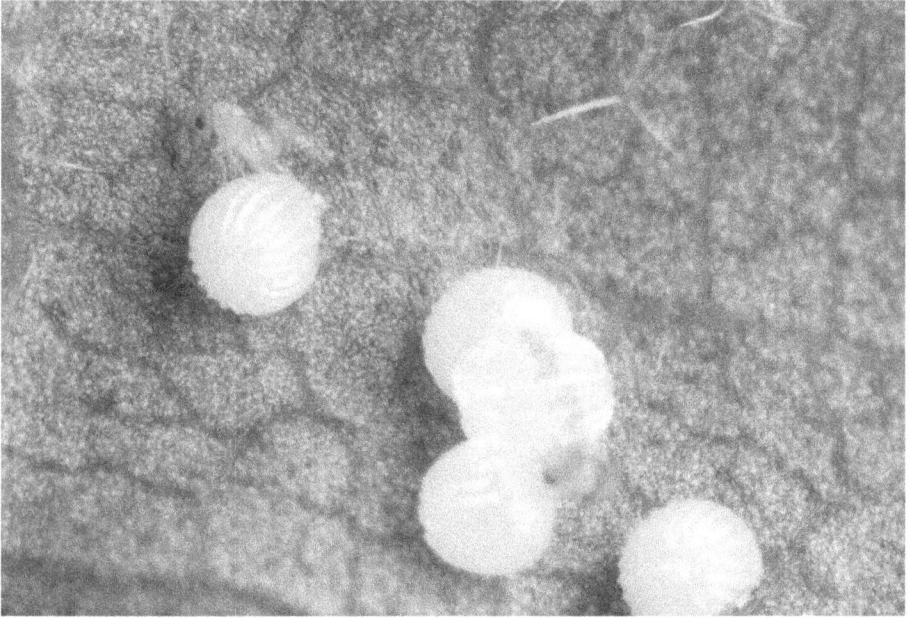

Trichogramma wasps are super-tiny. Here they are parasitizing insect eggs (they're the tiny orange insects). Wasps lay their eggs inside other insect eggs. Their developing larvae eat up the other insect larvae, then they pupate and emerge as adult wasps.
Photo by Peggy Greb, USDA Agricultural Research Service.

Gardeners can get ladybug larvae that eat live aphids, and praying mantises and pirate bugs that chow down on any insect pest that gets in their way.

Dragonflies and damselflies are good allies in the garden. These insects catch bugs while skimming through the air, and munch them while flying.

So the world of insects is filled with bug gore. E.B. White acknowledged this when he wrote Charlotte's Web (the only children's book to feature a pig, a rat, and a spider as heroes). Wilbur, when he meets Charlotte the spider, is heartbroken to find that his new friend is so bloodthirsty.

"I have to get my own living," Charlotte tells him. "I have to be sharp and clever, lest I go hungry."

Thus do a spider's wits – indeed, any beneficial insect's wits – makes for our gain in the organic garden.

Target the Right Bug

One day at the greenhouse, we were scratching our heads over what caused the holes in the leaves in several flats of plants. "Shoot, it looks like slugs to me," I said. I made that diagnosis mainly because the plants were in a cool, moist place, partially shaded, and I figured there were plenty of places for slugs to hide.

So Rhonda got the slug bait while I began to clean out the flat, looking for the slugs on the undersides of the leaves and the pots. Nothing. I was getting puzzled until, as I groomed the plants, a soft green caterpillar dropped into my hand. A few plants later, I found another caterpillar clinging to a stem. Hey, the slugs were framed; here's the real culprit!

The moral: it can be difficult to know which insect is making those holes in your leaves or toppling your plant, since pest insects stay out of sight, hiding under leaves or feeding after dark. There are plenty of other insects in your garden who take no interest in your leaves but simply happen to be hanging around the injured plant when you come outside. These poor innocent bugs end up being the target of your wrath, decimated in a cloud of poison. Then the pests come out of hiding, unscathed, and continue to use your plants as a salad bar.

Here, as always, observation is the key. Take care to investigate a plant for pests by cleaning up any debris around the plant, sifting through the mulch, and looking under leaves. When you know for certain what pest to target, you buy the right materials to take care of it, instead of wasting money buying, say, slug bait for a caterpillar problem.

Often, the problem is solved by handpicking the offending insects and squishing them. If the thought of handpicking makes you squeamish, I have heard of gardeners using a handheld vacuum cleaner to snap up bugs. No need to spend $11 on a quart of chemical you might spray only once or twice and then have to dispose of.

Slugs and snails are pretty easy to catch. You look for moist conditions and rainbow slime trails. Diatomaceous earth dusted on the ground around plants will kill slugs by razoring open their protective slime coating when they slide over it. Diatomaceous earth must be replaced after a rain.

A garden snail on its way to slime your tomatoes.

There's the old story about setting out shallow containers of beer for slugs, because they drink the beer, then fall in and die. This is not a foolproof plan. You need a lot of these containers, unfortunately, because the slugs can smell the beer from only a few feet away. And often they'll just take a few sips and then go about their business, though slightly toasted, one assumes.

Personally, I'd be more concerned about some raccoons sniffing out the containers, drinking the beer bait (and eating the beer slugs as snacks), and then having a big ol' drunken raccoon party in the yard until somebody calls the cops.

For a non-alcoholic slug bait, and to avoid having drunken raccoons trashing your yard, use three teaspoons of yeast and sugar dissolved in a cup of warm water, and put the dish right next to your slug-infested plants.

Vine borers can be a problem on squash vines. These are caterpillars that chew their way into the vine several inches above the ground, and kill off the squash when it blossoms. If you notice a sawdust-like substance on the stem, cut into the injured place and dig out the borer. Then pile up soil above the injured place. If you've caught the borer in time, the squash will send out new roots and survive.

Another insect to look out for is the large, dark-brown squash bug. Try laying out scrap boards next to your plants. Every morning, turn the boards over and smash the squash bugs you find there. You can also find and smash the rafts of copper-colored egg clusters on leaves.

Spending a little time investigating your problem plants always pays off.

A GALLERY OF BENEFICIAL INSECTS

A busy, buzzy bee

I love beneficial insects, but one thing that bothers me about the moniker "beneficial" is that an insect has to be helpful to be valued. There are millions of insects that are neutral on the scale between "beneficial" and "destructive" that still deserve to share this world with us and are interesting to watch go about their work. And those insects might be beneficial (or destructive) in ways we haven't figured out yet.

All that being said, inviting beneficial insects to your garden is a solid move on your part. In a world full of insecticides and chemicals and dwindling natural resources, a garden that invites the natural world in also creates an ecosystem where the helpful insects can live by snacking on the harmful insects and keep things in balance. And that's a world we can all live with.

Damselflies

*A damselfly at rest. Image by **Brett Hondow from Pixabay***

Odonata – Dragonfly family

I was sitting on the ground one day, pulling weeds, when I saw something odd. A little damselfly, hovering in flight, kept banging her head against the leaves of the small weeds nearby. Curious, I watched her for a while, trying to figure out why she was doing this. Then I noticed that the weeds had whiteflies, and the damselfly was picking the whiteflies off the backs of the leaves, one by one, to eat.

Damselflies are cousins to the dragonfly. All Odonata species develop in freshwater habitats or water-filled cavities. They lay their eggs in the water, and within one to three weeks, they hatch into a tiny naiad with a long body, long legs, and three leaf-shaped gills on its tail. (Dragonfly naiads are larger and lack the leaf-shaped gills.) They live in the watery world as a nymph, molting as it grows, eating every small creature it can get its mandibles around – mosquito larvae, small water bugs and spiders, and any other arthropods that get in its way.

Finally it climbs out of the water and lets the water evaporate off its exoskeleton. Then it molts, crawling out in dragonfly form. Once its wings have expanded, it flies away, gunning for flies and other bugs because after all that work, it's super-hungry.

Members of this family have their wing muscles attached directly to their wings (apparently other insect orders have their wing muscles attached to their thorax), so they can move each wing independently of each other, which allows them an insane amount of control in flight. They can hover, fly backwards, and do all kinds of stunts. So with superior flight skills, they can grab bugs out of mid-air using their hairy hind legs, and they can eat a snack in flight.

Dragonflies are stronger fliers than damselflies and are more likely to do the cool flying feats, and hawk bugs out of the air. I like meeting a dragonfly that's chilling on a fence post, looking all around with those big eyes. These insects come in many different colors – electric blues, shiny greens, flashy reds and bronzes. They are beautiful insects, and I wish they got the same press that butterflies do.

Damselflies are equal-opportunity bug eaters. If you're a little bug in front of a damselfly, you're going to be toast. Adult damselflies will hunt in groups to target termites, ant colonies, or a swarm of mosquitoes.

Even their young, the nymphs, are voracious eaters. Though the nymphs live in the water, they'll climb up other plants to find and eat insects.

Encourage dragonflies and their allies by having a small pool or a boggy water feature in your yard. The best way to keep damselflies and dragonflies coming around, as well as many other species, is by using fewer insecticides and having lakes and ponds and creeks with wild areas so these creatures can grow and develop.

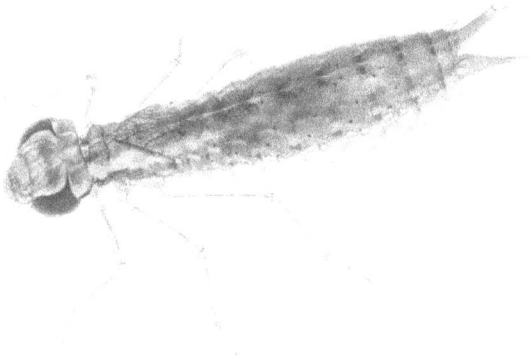

Dragonfly nymph (Anisoptera), also known as a dragonfly larva, photographed at Gavins Point National Fish Hatchery in Yankton, South Dakota. Photo: Sam Stukel/USFWS

Note: In some parts of Europe, dragonflies can transmit a parasitic flatworm to poultry. If a chicken catches and eats a dragonfly that is infected by flatworms, the pathogens infect the chicken's intestines. I don't recommend standing out in the chicken pen swatting away any dragonflies that are flying near the chickies, but I'm mentioning this so our European readers are aware.

Lacewings

A lacewing. Image by Frauke Feind from Pixabay

Neuroptera – Nervewing family

Lacewings (*Chrysoperla* spp.) are delicate-looking insects that are a real gem in the garden. They look like a skinny green lady in a gigantic shimmering dress, which is actually its two pairs of wide wings. Lacewings are wonderful predatory insects to have in the garden. The adult lacewings eat nectar, pollen, and honeydew, a very vegan existence; but the larvae are the complete opposite, and they're out to eat anything they encounter – aphids, lace bugs, caterpillar beetle larvae, insect eggs, mites, and sometimes each other. Like the honey badger, the lacewing larvae *just don't care.*

Lacewing larvae are also called aphid lions because they go after aphids like nobody's business. They grab aphids with their large mandibles, inject venom to paralyze the small insects, and then sucks out the juices. If you have aphids, these tiny lions will help clean them up.

Lacewing eggs are tiny white eggs, only a little bigger than a comma, and are attached to a leaf by a long stalk that holds them aloft.

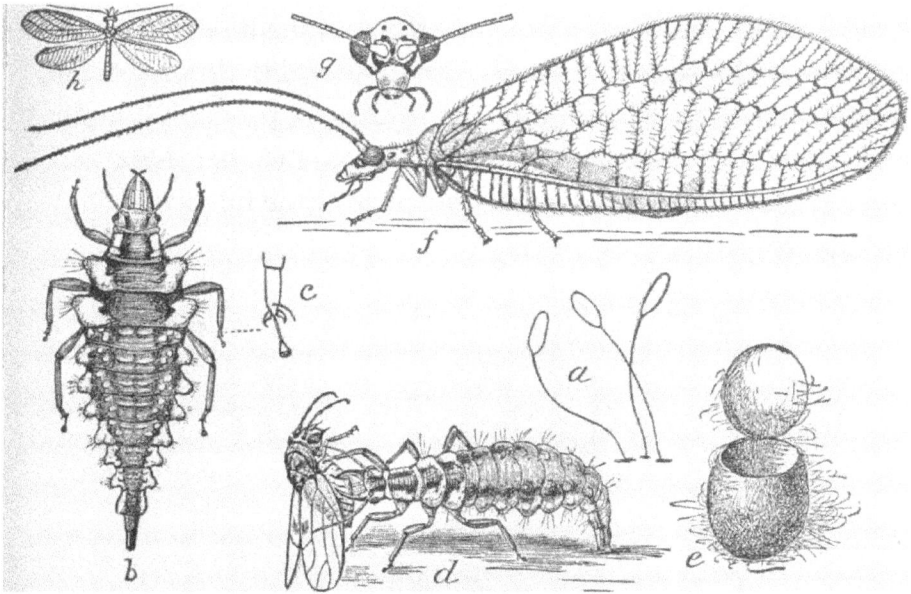

A: Lacewing eggs on the stems. B: A lacewing larva. C: Its foot. D: A lacewing larvae eating a hapless aphid for lunch. E: The pupal cocoon after the lacewing has left it. F: Side view of the pretty lacewing wings. G: Its face. H: The adult lacewing as it looks in flight.

The lacewing nymph that hatches out of the egg are usually a yellowish or red color, and they look like tiny alligators – and they are voracious predators. They'll even bite humans! I'm ashamed to admit that I used to squash them because those little jerks bit me – but I'm changing my tune now that I know what lacewing larvae look like. But they'd better not bite me.

You probably know about how ants will guard aphids on a plant, milking the aphids of honeydew and fighting off anything that threatens their "herd." Lacewing larvae trying to nab the aphids for lunch would be grabbed by the ants and thrown off the plant.

Some larvae have figured out a neat trick to bypass the ants. These larvae would cover themselves in debris – bits of lichen, dead aphids ,and other leftover parts from the insects they'd devoured – like a barbarian warrior. Under this cloak of death, they're able to sneak past the ants, who were probably thinking, "Oh, here is a walking pile of dead aphids, this seems legit." Then the debris-covered larvae is able to wade into a sea of aphids and happily eat its fill.

A green lacewing larva dining on whitefly nymphs. Photo by Jack Dykinga, USDA Agricultural Research Service.

You can buy lacewing eggs if you want to turn a lot of aphid lions loose in the garden. They will be shipped very close to hatching, and in a few days the larvae will start to emerge. They will be very tiny. They'll need to be released quickly before they start to eat each other. If you can't release them immediately, slow them down by putting them in the fridge, NOT THE FREEZER because the freezer will kill them. Don't leave them in the fridge for more than 48 hours.

Take the tiny larvae outside and release them in the cool of the morning or in late afternoon in various places so they can spread out and attack pest species and not each other.

Ladybugs

A ladybug chilling on a cornflower petal. Image by Ralf Kunze from Pixabay.

Coleoptera – Beetle family

I'm giving the family name of this particular beetle, instead of listing an individual ladybug species, because there are so many ladybeetle species throughout the world, even in Missouri. The ladybug (Coccinellidae family), a beneficial insect, is often red or brown, and often spotted (but not always).

Ladybugs are big sellers in organic gardening catalogs – kids love them, and so do grown-ups. When you release a bag of ladybugs in the garden, the adults will eat bug pests, but eventually they'll lay eggs and then leave. After a while you'll see tiny black and orange larvae that look like miniature Komodo dragons on your garden plants. These are ladybug larvae, and they will eat aphids and other insect pests.

The ladybugs often lay their orange or yellow eggs near aphid colonies. The ladybug larvae, which is black with orange markings, is a predator, voraciously attacking aphids, scale insects, mealybugs, beetle eggs, and anything else that moves, like a tiny black and orange alligator.

A ladybug larvae (the black and orange dragonish insect) picking up a little aphid lunch. Image by u_3heuehh9 from Pixabay.

The larva grows larger and larger, shedding its skin every time it does. It is an effective and hungry predator, often eating its own weight in aphids every day. Once it has grown, the larva pupates to turn into an adult ladybeetle.

The adult ladybugs are just as predatory, but they also dine on pollen, nectar, and honeydew.

One drawback to ladybugs is that, like the old poem says, they will "fly away home." The larvae are guaranteed to stay on your plants, but once they metamorphose and develop wings, you might be out of luck.

However, growing flowers around the garden, whether vegetable or flower, is a great way to encourage beneficial insects – including ladybugs – to make their homes here. Sweet alyssum is an especially good choice to attract ladybugs and other beneficial insects.

There are a few lady beetles that are not beneficial. The Mexican bean beetle and the squash beetle, which are yellow-orange with black spots, are destructive pests in the garden. The Asian ladybug is a larger ladybug that is very helpful in the garden but tends to congregate in old houses in fall and winter. Also, they bite! But overall, everybody loves ladybugs.

Wheel Bugs and Assassin Bugs

A wheelbug on a small sunflower. Note the cogwheel-shaped ridge on its upper abdomen which gives the wheelbug its name. Photo by Jeffrey Hamilton on Unsplash.

Hemiptera – True bugs

The above photo shows the wheelbug, a type of assassin bug (*Arilus cristatus*). Wheelbugs are in the Reduviidae family, or the Assassin Bug family.

These particular insects are fairly large and have a prominent proboscis, which is the straw-like mouthpart that these insects feed through. (This proboscis is also a prominent part of the Hemiptera family, which largely feeds through this straw –

whether they drink liquid from plant leaves or liquid from other bugs.)

Wheelbugs will ambush their prey, jab them with their proboscis, and inject salivary enzymes into it to break down their innards for easy sipping. The wheelbug then carries the poor insect around on its proboscis like it's a Slurpee from the 7-11.

I turned this picture sideways so you can see the assassin bug better. Image by eszobel from Pixabay.

Assassin bugs, seen above on an ear of corn after having made a Slurpee out of a weevil, are great helpers in the garden. They will catch beneficial insects as well as destructive insects. But keep in mind that if you're a predatory insect, everything you see looks like lunch.

Wheelbugs are great to have in the garden, but holy crap, don't try to make friends with them. Don't pick a wheelbug up because they will stab you with that proboscis, and that thing hurts like hell. They are called "assassin bugs" for a reason! One landed on my forehead one time. I reached up to brush it off, and it stabbed a knife into my scalp. Man, that hurt!

However, these insects are very good at taking out caterpillars, beetles, sawflies, and aphids.

Other related bugs are ambush bugs. These bugs lurk around flowers, and they have big, muscular forelegs, which they use to capture flies, bees, and other insects like a lion springing out of the grass to capture its prey.

Praying Mantis

Praying mantises are a lot of fun until they jump on your head. Image
by Josch13 from Pixabay.

Mantidae – Mantid family

Now these guys are great, and they don't even stab you with their little knives.

Praying mantises (*Mantis religiosa*) are often green or brown and busily eat other insects (and sometimes each other). Also they're fun to play with. If they're sitting on your hand, they'll start swaying back and forth hypnotically, and if you let them do this long enough they will probably jump on your head.

If you put a mantid on your computer screen and start moving the cursor around on the screen, sometimes the mantid will chase it and try to catch it. Like I said, these guys are a lot of fun.

Praying mantises are like tiny lions, devouring anything that crosses their paths. They'll often be sitting in a garden with the remains of some insect clutched in its arms, chomping away.

Don't get them mixed up with stick insects, which resemble mantises, except stick insects look like green or brown sticks, and they don't have the curious, triangular, big-eyed head that a mantis has. Also, stick insects eat leaves and can be a pest insect.

Mantis egg cases, actually a form of solidified foam, look like a brown packing peanut stuck to a twig. The young mantids hatch from the eggs, and they sometimes end up devouring each other, so eager they are to get something to eat.

Mantids are sold through garden supply companies to be released in your garden, if you want to build a mantis paradise in your backyard.

Big-Eyed Bugs (Geocoris spp.)

*Having glued a hapless whitefly to a leaf, the big-eyed bug can devour its prey at its leisure. Photo by **Jack Dykinga**, USDA Agricultural Research Service.*

Hemiptera – True bugs

Big-eyed bugs (*Geocoris* spp.) are small, about 4 mm, or about 1/8 to ¼ of an inch long, with that familiar "x" in the wings that categorize many insects in the Hemiptera (true bug) family. These are predatory insects (the bulging eyes on the sides of their head gives them a wide field of vision to spot their prey), and they go after a wide range of food sources, including small caterpillars and their eggs, flea beetles, mites, aphids, whiteflies, and other small prey. Big-eyed bugs will also feed on seeds (another name for them is "Seed bugs") or suck plant juices, though not in a way that injures plants.

Like the assassin bug and most others in the Hemiptera order, these little guys feed through their proboscis. They jab this into their meal, inject digestive enzymes, and then they suck up the partially-digested innards of the unfortunate insect. They are small but very helpful.

These insects closely resemble cinch bugs, which are a real pest in the garden. However, big-eyed bugs have the gigantic eyes and are small and stout, while cinch bugs have small eyes and have narrower bodies.

Wasps

A tiny parasitic wasp, Trissolcus euschisti, emerges from a stink bug egg. Its mama inserted this wasp's egg inside the stink bug's egg, where the wasp larvae hatched and then ate the stink bug larvae. Pretty sneaky! Photo by Elijah Talamas, USDA Agricultural Research Service.

Hymenoptera – Sawflies, wasps, bees, ant family

I wasn't a fan of wasps in my younger days. When I was a kid, they'd get into the house and fly around, and I would be the first to grab the scissors and cut them into twitching bits. I was so morbid. But now all that has changed. I like wasps!

I grew to like wasps when I was in college. I had my writing desk set against a window with wasp's nest in it, and since I never got around to taking the nest down, I ended up getting a close-up view of how wasps live, and began to enjoy watching them.

The part I liked best was feeding time, when one of the wasps would squeeze in under the outside pane. She had in her mouth (i.e. mandibles) a round ball of some greenish stuff, which I

47

discovered later were chewed-up caterpillars. When she gained her feet, she buzzed drunkenly up to the nest, then would spend some time turning the ball of green meat in her front legs, chewing and softening it all the while, before tearing off pieces and feeding them to the grubs. When mealtime was over, she always washed herself carefully, like a cat.

Most people are not too crazy about wasps, and with good reason. I remember when my fellow Park and Rec workers took out a wasp's nest at Krug Park after wasps stung a girl twice. But if the nest is out of the way, wasps are beneficial to your garden.

Wasps catch the bugs that wreak havoc with your plants. Wasps feed a lot of bug meat to their larvae from aphids and caterpillars they catch. Wasps also help pollinate flowers.

I like wasps in general, though I will never ever ever tangle with any wasp with yellow markings, as I am a "safety first" kind of gal. Image by Willfried Wende from Pixabay.

Different varieties of wasps target different kinds of insects. For example, mud daubers will capture small spiders for its young, and cicada killers, which are very large wasps, eat cicadas.

Other varieties of wasps are beneficial because they lay their eggs in living caterpillars or other insects. The eggs hatch and the larvae devour the poor caterpillar from within.

Robert Frost, the Vermont poet, allowed white-faced hornets into his kitchen because they were so good at catching flies. Once in a while, though, the hornets would go after things that looked like flies, including huckleberries.

"He stooped and struck a little huckleberry the way a player curls around a football …. The huckleberry rolled him on his head," Frost wrote in his poem, "The White-Faced Hornet."

Sometimes, a wasp will build her nest too close for comfort, such as when they build one in your garden hose caddy. If you have to kill wasps, find a spray that allows you to stand far away when you're spraying it, and keep your eyes open for wasps defending their home. Always be ready to run!

If you need to knock down a wasp nest, do it early in the morning when the temperature is chilly. Wasps are sluggish in the cold and will be slower to react.

Cicada Killers

How a cicada killer brings home the bacon – er, cicada

Hymenoptera – Sawflies, wasps, bees, ant family

One hot summer afternoon as I worked in the garden, I heard a small thwack in the air behind me. A cicada screeched. I turned. A big wasp with a cicada in its grip spiraled to earth, the cicada whining the whole way down. The wasp, which was over an inch long, was a cicada killer (*Sphecius speciosus*).

One would think that the biggest wasp in North America, one that can catch a cicada twice its weight, would be tough stuff. But no: this gigantic wasp is a wimp toward anything not a cicada. If you are crazy enough to get a cicada killer to sting you, you would have to sit on the poor guy. Even if you disturb its nesting burrows, all it does is buzz loudly.

The wasp kept stinging the cicada it had caught until the cicada stopped fighting. It was paralyzed, not dead. Then the wasp, using its mandibles and legs, flipped the cicada onto its back, wings down. The wasp stepped over the cicada, picked it up with its middle legs, and dragged it forward. The cicada's wings worked as a sled. When they got snarled in a clump of grass or behind a twig, the wasp would buzz and carry its load over the obstacle.

The wasp started dragging the cicada toward me, so I stepped out of her way. The wasp turned toward me again. I backed up, but she kept coming. I thought, *Either she really, **really** wants to be my friend, or she thinks I'm a tree.* I backed up against an oak, waited until she was close, then stepped aside. This time, she kept going toward the oak.

If a cicada she's carrying is heavy, she will carry it up a tree, then fly off a branch. This will give her enough altitude to fly the cicada back to the burrow she's dug for her eggs.

Cicada killers are digger wasps, ground-dwellers, and they'll dig a burrow two feet deep and three feet long to lay their eggs. At its end are several round rooms: the nest-cells. She'll drag the paralyzed cicada into a nest-cell, lay an egg on it, then seal the cell. Now this is a little gross: When the wasp grubs hatch out, they eat the still-living cicada. In May or June, the grubs metamorphose into wasps, dig their way out, and fly away.

I get such a kick out of watching cicada killers in the garden. There was a cicada killer that used to hang around the rose garden, pouncing on little pieces of mulch that (I guess) she kept mistaking for cicadas. She apparently needs a pair of glasses. She was a big, mean-looking wasp, almost two inches long with black and yellow stripes. I nearly stepped on her several times but she never made any threatening moves. She ignored me when I walked through her territory – thank goodness.

A cicada killer on her little perch, looking out for trouble (or lunch). Image by Bruce Emmerling from Pixabay

One afternoon I saw a cicada killer sitting on a rose, turning her head left and right as if looking around. She would zip out and chase away a fly, then come back. She chased away a butterfly and a grasshopper. She flew at a spider in her web and it scrambled into a hole. She tried to chase me. I didn't run, so she sat back down on the rose.

Another time I saw a cicada killer perched on a short stone column, sallying forth to chase various things away. While he was chasing a fly, I held up a finger right next to his perching spot and a little bit higher. The cicada killer came back and perched on top of my finger. It tickled.

In the rose garden, a cicada killer dug a nest-burrow in my underground hose bib. When I took the lid off, she'd shoot straight up in the air and hover in front of my face. I just told her "Sorry!" and hooked up the hose. Then she'd buzz away.

I always miss the cicada killers when fall arrives. However, I'm always happy to see them the next summer.

Leaf-Cutter Bees

Cuckoo leafcutter bee (Coelioxys spp.). Public domain image by Alejandro Santillana produced as part of the "Insects Unlocked" project at the University of Texas at Austin.

Hymenoptera – Sawflies, wasps, bees, ant family

Leaf-cutter bees, a pollinator that's native to the United States, will make an appearance in summer, cutting neat circles in your rose leaves. These leaf bits are used inside their nest cells, which they build inside the stems of soft-pithed woody plants, such as roses or willows, or soft, rotted wood. If you have trimmed your roses back, sometimes you'll see a small hole in the pith with a little frass or sawdust bits around it. This is likely the work of the leaf-cutter bee, though it could also be from a carpenter bee or a small wasp. The leaf-cutter bee tunnels its way into the pith, and then use the cut leaf pieces to create nest cells. Each cell is provisioned with pollen and nectar, and once this is finished, she lays an egg, closes up the cell, and buzzes out to fetch some more leaf clippings to start a new cell.

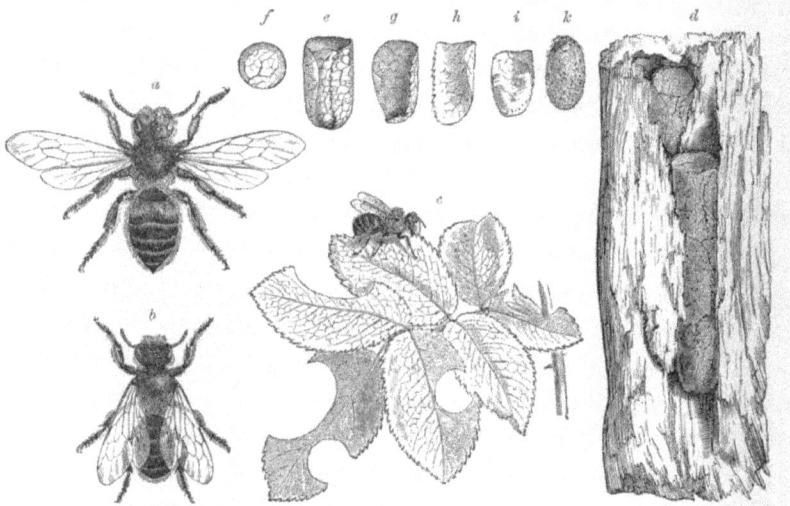

COMMON LEAF-CUTTER BEE. *a, b*, Female and male (enlarged) ; *c*, Rose-leaves with several pieces clipped out and a bee at work ; *d*, Nest in a willow stem ; *e*, A single cell ; *f*, The lid of same ; *g-h*, Pieces of leaf ; *i-k*, Side pieces.

a, b: Female and male leaf-cutter bees; c: Rose leaves with several pieces clipped and a bee at work; d: Bee nest in a willow stem; f: The lid of the nest; g-h: Pieces of leaf; i-k: Side pieces

There's not much you can do about leafcutter bees, because once you find their damage, they've already moved on. Though the small bees and wasps that burrow into your rose pith generally don't damage the rose (they seldom cut into the cambium, or growing layer), they aren't a concern. However, Charles, my late rosarian friend, would seal exposed rose piths with a small drop of white glue to keep the small bees out.

Since they're pollinators, and the honeybees need all the pollinating help they can get, spare the leaf-cutter bees.

A ROUGE'S GALLERY OF GARDEN PESTS

Mealybugs

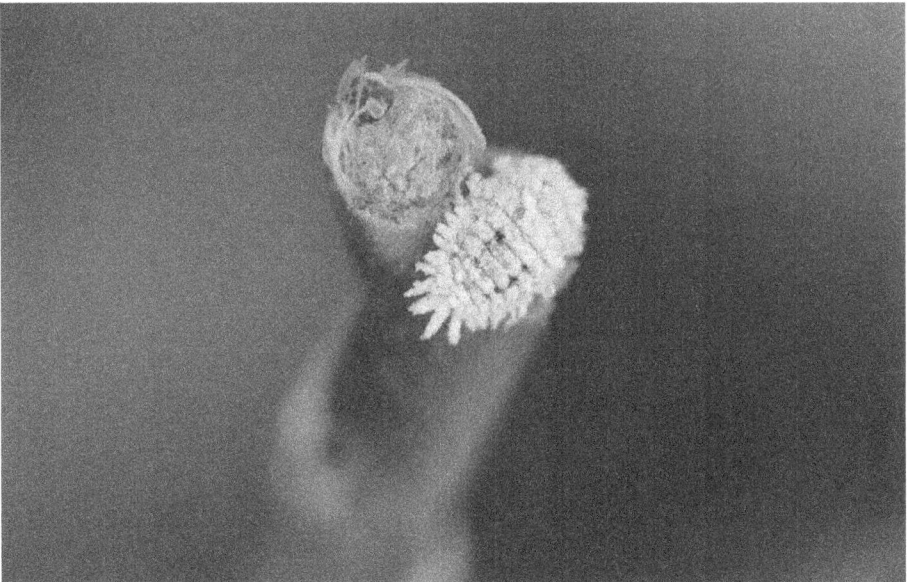

Mealybug adult. Image by Sandeep Handa from Pixabay

Hemiptera – True bugs

Mealybugs look like white bits of cotton tucked into plant crevices. The cotton on their bodies is actually wax.

These insects afflict indoor and outdoor plants. Mealybugs are generally wingless, whitish, oval bugs about 3 millimeters long, and look like tiny bits of waxy fluff. Some have long white "tails". Newly hatched mealybugs, called nymphs, are flat, oval, and yellow. Once females mature, they stick themselves to the plant, cover themselves with a powdery wax layer that repels water (and insecticides), and suck plant juices.

Mealybugs feed by sticking their threadlike mouthparts – the proboscis – into the plant, inject a little of their saliva into the plant to liquefy the cells in that area, then they suck out the plant juices. Naturally, when a bunch of mealybugs do this, the plant becomes stunted, or wilts, or dies back.

Mealybugs multiply easily and can be hard to kill, and by sheer numbers can overwhelm the plant. They also excrete honeydew, a sugary substance, and in a plant with lots of mealybugs, the honeydew they leave behind is infected by sooty molds, turning the plant a sooty black color.

Male mealybugs resemble whiteflies or tiny white gnats when mature. They have wings and can fly around. However, females are wingless, and have to be transported to new host plants to infect them. They can crawl short distances, but more often, the nymphs can be blown by the wind to a new plant to infect, or can be picked up as stowaways on the feet of birds. Most often, an infected plant is set down next to an uninfected plant, leaves touching, and the mealybug nymphs gallop across the gap like wild horses.

When they first arrive on a new plant, mealybugs tend to wedge themselves into the crevices, generally in those tiny spaces where a leaf joins the plant stem. But then a female makes an ovisac, which looks like a waxy bit of cotton, and lays some eggs in it. Once that first batch of eggs hatch, that's when the infestation starts. Other females join in the egg-laying spree, and then you have a mealybug crisis on your hands.

Some mealybug species, such as the Madeira mealybug, can have up to five or six generations every year. Imagine multiplying so fast that you could live on the same plant as your great-great-great-great-great-grandchildren in the space of a year! Females can lay, on average, 300 to 400 eggs in their lifetime. Sometime they can lay eggs that immediately hatch.

So for all these reasons – the rapid-fire egg laying, the many generations in a single year, the relative waterproof-ness of the insects and the wooly egg sacks – it's a challenge to stop a mealybug infestation. Outside, predators such as green lacewing larvae and ladybugs can attack mealybugs and bring their numbers down (except when ants protect the mealybugs, farming them for their honeydew). But inside, without predator insects, control is even more of a challenge.

Use several points of attacks in getting rid of infestations. Isolate the afflicted plants. Spray neem oil on the places where the mealybugs are congregating. You can also spray them with insecticidal soap or pyrethrin.

Also, dip a cotton swab into rubbing alcohol and touch it to each mealybug you see. The rubbing alcohol will eat through the waxy coating that protects the insect, drying it out and killing it.

But if you spray them, you should also squish them. Mealybugs can develop resistance to different insecticides – but they'll *never* develop resistance against being squished.

Check the bottoms of the leaves and stems for mealybugs and their nymphs and squish them wherever you see them. Use a toothpick to squish any mealybugs that have packed themselves into the crevices of your leaves and any other tight places where your fingers can't reach.

I recommend squishing or otherwise wiping out any insect pests when possible if it doesn't make you feel ill.

I have a mealybug story that helps to illustrate why it's best to use a hands-on approach to pest control.

I had a mandevilla in the greenhouse that had a huge mealybug problem. I sprayed the plant with insecticide until the leaves were dripping. The mealybugs were still there. I put a systemic insecticide around the roots of the plant and watered it in. The mealybugs didn't care.

So I just started squishing the mealybugs with my fingers, a gross job. At that point, I didn't care. I searched them out and squashed them where they were cuddled up around buds, in the cracks of the plant, and under the leaves. I even found some on the roots just under the soil. I squished those and added a little extra potting soil. I checked the plant every other day and squished every mealybug I could find. After a while, I stopped finding them altogether. Then I fertilized the plant, and the mandevilla put out leaves like crazy and started blooming. Success!

Chemicals aren't a cure-all. Sometimes you just have to get your hands dirty with the plant to help it along. It's a good feeling when a plant you've been working with rights itself and perks up again.

Ants will defend mealybugs on plants as they do with aphids, since mealybugs also produce a lot of honeydew. I talk about dealing with this double-trouble threat in the aphid chapter.

You can also use insecticidal soap in a root drench for root mealybugs and springtails – some bugs that you don't see often, since they're hiding in the soil. (Springtails are tiny insects that spring into the air when you try to squish them. They're like fleas, except you find them on the soil, and they drink plant sap, not blood). Add 1 to 2 tablespoons of concentrated insecticidal soap to 1 quart of water and pour this solution on the soil. One drench each month should clear them up. However, it wouldn't hurt to take some of the soil off the tops of your plant's roots to check on the infestation.

Whiteflies

Whiteflies (specifically, Bemisia argentifolii) on the bottom of a leaf. These insects are about 1/16th of an inch long. You can also see in the photo their discarded pupa cases and a few eggs (in the upper part of the photo). Photo by Scott Bauer, USDA Agricultural Research Service.

Hemiptera – True bugs

It's crazy how many of these common insect pests are from the Hemiptera order. But this is one instance when the name of an order tells you what you can expect. When you see an insect from the Hemiptera order, you can expect a tiny bug sticking a straw into the plant cells and sucking up the juices, and multiplying faster than a math whiz in a counting contest.

Whiteflies are no different. You'll usually find them on the undersides of the leaves. The whitefly life cycle is egg, nymphs, pupa, and adult. The immature whitefly nymphs don't have wings and are called crawlers, and they look like tiny, waxy white scales on the bottoms of the leaves. They don't move much. The whiteflies use piercing-sucking mouthparts to suck the juices out of the plant leaf. In some places, they can spread the virus that causes tomato yellow leaf curl.

Despite their name, some forms of whiteflies are actually black or grey, but they generally have something of a flour-y look, as if they were dusted with flour. There are hundreds of different species of whiteflies afflicting many different kinds of plants.

You can tell a plant is badly infected when you touch the plant and a swarm of tiny whiteflies rise into the air from the leaves.

Like its annoying distant cousins, the mealybugs and aphids, whiteflies also excrete honeydew to get ants to protect them from predators.

Whiteflies larvae resemble mealybug nymphs. They crawl around until they find a place to settle down and feed, whereupon they stop crawling and attach themselves to the plant by its mouthparts (a proboscis). After a time, it changes inside its own skin, creating something like a pupa. Once its metamorphosis is complete, out pops the adult whitefly, complete with wings.

Like other members of the Hemiptera family, these insects can cause sooty mold to attack the plants they've dropped honeydew on.

Also, while the whiteflies are feeding on the plants, they inject saliva into the plant cells to break down the cell innards so they can easily suck them up. Naturally this is toxic to plants when whiteflies feed in large numbers.

Whiteflies can also transmit diseases and harmful viruses from plant to plant, such as tomato yellow leaf-curl begomovirus, a devastating viral disease.

When a plant is infected with whiteflies, immediately separate it from the others and wash it off with water with a little soap added. Gently rub the leaf bottoms with the soapy water to rub off whitefly larvae and strip them of their protective waxy coating. Spray the plant with water to further knock off any larvae, let the plant dry, then apply neem oil, insecticidal soap, or pyrethrins.

Repeat this regimen weekly. Put up yellow trap cards around your plants to catch any flying whiteflies.

Whiteflies quickly develop resistance to insecticides, so include mechanical means of control – the soapy water scrub and the yellow trap cards – for best results. Remove any yellowing or dying leaves and seal them in airtight bags so whiteflies don't hatch on these and repopulate your plants.

You can even vacuum the air around your plants if the infestation is that bad – or if you feel particularly frustrated. Start your vacuum, shake the plant, and suck the flies right out of the air. Take care not to vacuum up your plant, of course. You might hold your plant with one hand and hold the vacuum nozzle with the other, and try to keep them far enough apart to avoid damaging the plant, as you shake the plant and brush off the bottoms of the leaves. Dispose of the vacuum bag away from your plants, sealing up the dust so those little jerks can't escape.

Bring hummingbirds into the garden, as these little birds also eat whiteflies. It couldn't hurt!

Scale

White scale on a leaf. Photo by Mokkie.

Hemiptera – True bugs

Here is another Hemipteran. Scale are part of a diverse group of insects. Though they are related to aphids, whiteflies, and mealybugs and share certain characteristics, these bugs sport a wax covering that makes them resemble a fish scale.

Scale insects have a wide variety of coverings in a variety of shapes and colors. You can find white, waxy scales, as in the photo above. Others look like tiny brown water spots on the underside of leaves, or mussel shells, or bits of white fluff. Like aphids, some scales have characteristics of both sexes, and some can birth live young without fertilization.

The female lays eggs under her scale-like body, and then the nymphs, or crawlers, creep away to find a permanent location to sip plant juice, where they stick themselves to the plant by their mouth parts. They are stuck there for good, because adult scales seldom move.

Scales will cause yellowing leaves, distorted foliage, and eventually the leaves will fall off.

Scale insects are hard to detect when they're small, but honeydew – the sugar that these insects excrete – reveals their presence. The sticky, sugary honeydew on the leaves gives rise to sooty mold, which blackens the leaves, and this attracts ants, wasps, bees, and flies.

Controlling scale can be a challenge. On houseplants, you can dip a cotton swab in rubbing alcohol and touch it to the scales, which destroys the waxy coating and dries them out. It's time-consuming, and you'll need to keep doing this every couple of days, but it's very effective. Insecticides are not very effective against adult scales due to their waxy coating, though they do work against unprotected crawlers before they grow their protective shell.

On outdoor plants, applications of insecticidal soap or horticultural oil will be effective if you spray every six or seven days to keep catching new crawlers as they hatch out. (Horticultural oils should not be sprayed in temperatures over 85 degrees, or when the humidity is over 90%, as it may damage the leaves.) Horticultural oil is a fine-grade oil that coats the insects and blocks their spiracles to suffocate them. (Spiracles are a series of tiny holes in the insect's thorax and abdomen through which it breathes.)

Stink Bugs

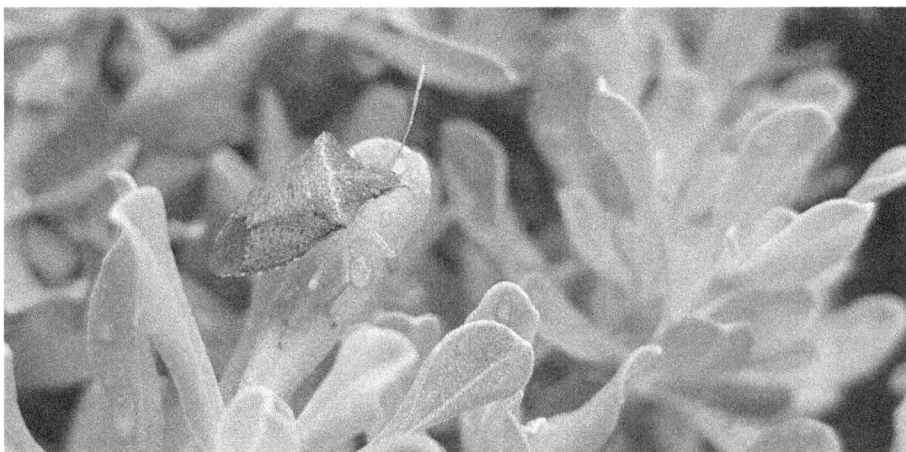

A stink bug. Image by Brett Hondow from Pixabay.

Hemiptera – True bugs

Stink bugs are about the size of a fingernail. They're shield-shaped, green or brown, and can blend in very well with the plant or ground. I learned about stink bugs when I was a kid walking in the woods. Sometimes I'd sit down in the leaves to take a rest, and then I'd smell a pungent odor from the stink bugs I disturbed. The smell is much, much worse if you put your hand in the leaves and accidentally squish one.

What kind of damage does a stink bug do, besides offending the nose of young nature enthusiasts.

Stink bugs stick their piercing-sucking mouthparts under the skin of your fruits or leaves and inject an enzyme that liquefies the plant cells. Then they suck up the liquid. These bugs (they are actually true bugs) leave dark pinprick spots on the fruits and leaves, or odd knots, or a light stippling on the fruit's surface. I've seen those odd stipples on my tomatoes, and didn't know what caused it until I wrote this chapter. One mystery solved.

A stink bug infestation can cause severe damage to the fruits. When you have a lot of stink bugs sticking their mouthparts into your tomatoes, now and then you'll get a mouthpart that has some yeast on it from heaven knows where. Once the yeast (or some other microbe) gets inside the tomato, the fruit is infected.

Predatory insects, like ladybugs, assassin bugs, and wheel bugs will attack stink bugs. Chickens may be helpful in controlling them.

You can control the insects by squishing them (if you can put up with the stink), dumping them in a bucket of soapy water, and spot-spraying infestations. Use several control methods at the same time – this will usually bring home the bacon.

Fungus Gnats

Fungus gnat. Photo by Andy Murray.

Diptera – Flies

Fungus gnats hang around potted plants, but the adults of the species don't harm them. All that the gnats eat are liquids like water and plant nectar – but they're annoying as heck, flying around all over the place. They generally fly straight at my eyes, just to add insult to injury.

Yellow sticky traps hung near soil level will catch many fungus gnats, and the occasional whitefly. You can find sticky traps at pet stores or farm-supply stores. But catching the adults is only half the battle, because you also need to kill the real troublemakers: The larvae that are developing underneath the soil and eating the plant roots.

Fungus gnat larvae consume fungi and decaying matter, but they also eat your plant's root hairs and feeder roots. If you are trying to start geranium cuttings, you might sometimes pull up the cuttings and find the larvae have been nibbling all around their cut edges (and sometimes you can even catch a larva wriggling around the base of the cutting).

To test for fungus gnat larvae, push a cube of potato a little way into the soil. After a week, lift it out and check the cube and the soil underneath for tiny, white worm-like critters.

To discourage fungus gnat larvae, allow the plants to dry between waterings. The larvae don't like dry conditions.

Drench the soil with insecticidal soap by making a solution of insecticidal soap and pouring it on the soil. This will also kill springtails, an added bonus. Follow label directions when doing this.

Mosquito bits, or *Bacillus thuringiensis israelensis* (also called BTI), is very effective against larvae. Using a pencil, poke the bits down into the plant's soil about four inches, then water. The larvae will eat the BTI and perish. These will need to be reapplied after two weeks. Since BTI targets mosquitoes, these are particularly effective against fungus gnats (which are in the same order – Diptera, the fly family).

What's easiest is to put a little chunk of mosquito dunk into my watering pot, fill it with water, and let it sit for a day. Then water. The BT in the water will gradually kill off the larvae until soon you have hardly any.

I've also had great success by bringing spiders over to the infested plant. If you're okay with spiders, they're very helpful to have around in general.

If the infestation is bad, take the plant out of the pot, shake the dirt gently off the roots, and repot it in a fresh container of brand-new potting soil. Get the old soil out of the house before any more gnats hatch from it!

Controlling these little pests is mainly a matter of cleaning up any place where their larvae can wriggle. Once the larvae are gone, the flies will be gone too, and your peace of mind will be restored.

Flea Beetles

A flea beetle, much enlarged. Image by Brett Hondow from Pixabay.

Coleoptera – Beetles

These are tiny beetles, a little bigger than actual fleas, but they also jump – hence the name. They multiply like crazy on tomatoes as well as eggplants and other plants, chewing small holes in the leaves, and in no time at all the plant is riddled with holes. Flea beetles can also defoliate and kill plants when they really get out of hand.

The flea beetles overwinter in the soil or garden debris, and when spring warms them up, they come out and eat new growth on their host plants. In late spring or early summer, they lay tiny white eggs in the cracks of the soil around the plants, and the tiny white larvae feed on plant roots. After the larvae pupate, the beetles emerge and head right up into your plant.

Flea beetles overwhelm the plant (and the gardener) through sheer numbers. But as with any insect pest, if you fight the war on several fronts, you will be victorious.

Golden flea beetle (Systena spp.) Washington, DC. Note the muscular hind leg on the beetle. Flea beetles come in many different colors and shapes, but they are very tiny, and they can really jump! Photo by Katja Schulz.

Blasting the plant with water will knock flea beetles off the leaves. Follow this up with an application of neem oil or Azatrol EC, which of a pumped-up version of neem oil. Insecticidal soap also works, as well as kaolin clay, which is sold as Surround WP. The watered-down clay coats the plant while allowing the plant to respirate and photosynthesize – actually, kaolin clay seems to enhance photosynthesis – and keeps the beetles off without harming the helpful insects you need in your garden.

One control method that occurred to me while reading about the life cycle of the flea beetle is this: clear away the mulch around the plant and put black plastic around the plant. Tuck it in tight around the stem and overlap the seams, to keep the beetles from laying eggs, and to prevent the newly-hatched adults from getting out of the soil and onto your plant. I am ordinarily very much for mulch. But if flea beetles have been a problem in your garden for a while, this solution might cut their numbers down and give your other control measures have a chance to work.

Another control method is to put floating row covers over your seedlings right after you plant them to keep adult flea beetles from migrating onto your new crops. This should keep the damage down and help your additional control methods as well. (Since these are beetles, they can fly in – as well as jump.)

Tomato Hornworms

Tomato hornworm eating a tomato down to the bare stem. Image by Margaret Martin from Pixabay

Lepidoptera – Butterflies and Moths

Tomato hornworms are gigantic green Caterpillars of Enormous Size, and they range from about the size of your little finger to the size of a pickup truck – well, maybe not that big, but close enough. These caterpillars will eat your tomato leaves to skinny, stunted skeletons and drop big caterpillar poop everywhere. They are ugly as sin. I generally will squash any insect between my fingers, but I will not squash these. I don't like to step on them, either, because they pack an incredible amount of green goo in their innards.

Some people say that chickens like them. This may or may not be true. I used to offer tobacco hornworms to the geese and ducks at the city rose garden, and they wanted nothing to do with them. When I found these on my nicotiana plants, I'd grab them and heave them as hard as I could toward the lagoon, watching with satisfaction as that damn caterpillar soared like a bird toward the water and the rocks.

Handpicking these caterpillars is the best way to go. You would think these would be easy to find, but they can be deceptively plant-like, even for their enormous size. If you notice some of your tomato leaves have been defoliated, immediately look for caterpillar poop, because these caterpillars are walking intestinal tracts. If you see some, go straight up on the plant until you find the caterpillar from whence it came. Then drop the caterpillars in a bucket of soapy water, or mail them to your congressman (if he's a jerk).

T they're doing.

You can kill off the caterpillars at the end of the season. When your tomato plants are finished and the frosts are here, pull up the plants, then till the soil to expose and kill burrowing caterpillars and pupae.

A sphinx moth. These are actually very pretty. Picture by Brett Hondow from Pixabay

The adult form of the hornworm caterpillars is the sphinx moth, or the hummingbird moth. I like these moths a lot, and you see them at twilight, humming in front of the petunias and sipping from the blossoms. But I have no love for the tomato hornworm.

Aphids

Aphids (aka greenflies) on a rose. Image by PollyDot from Pixabay

Hemiptera – True bugs

Aphids are small bugs, but they can pack a wallop because they reproduce so quickly and can take over your plants fast. Aphids, sometimes called plant lice or greenflies, are tiny sap-sucking insects that appear on soft stems and new foliage. Generally they are green but sometimes you'll find black or wooly white aphids taking over your plants.

Aphids reproduce by laying eggs AND through giving live birth AND they don't need to sexually reproduce, because females can create baby aphids on their own! Great news, right? or not.

The green, pear-shaped bugs have no exoskeleton to protect them from predators, which means that any ladybug or lacewing can find them and eat them up. So how do aphids survive so successfully?

The answer: Ants protect them. When you stroke an aphid's back, the aphid secretes honeydew – that is, they excrete a little drop of sugar. Ants have figured this out, so when the ants discover that aphids have taken over a plant, the ants will organize and protect the aphids. The ants stroke the aphids' backs to get honeydew, then take the honeydew back to the anthill. Ants will "farm" aphids the way a farmer raises dairy cattle. Some species of ants will even carry aphid eggs back to the anthill in fall so the eggs have a safe, warm place to hatch, then put the newly-hatched aphids back onto the plants in spring. Ants will pick up aphids in their mandibles and carry them around (unlike humans and their cows).

Under the ants' protection, aphid thrive, sipping plant juices through their piercing-sucking mouthparts and leaving honeydew all over the leaves of the host plant, which causes sooty mold fungus to attack the plant.

Ants will fight for possession of the aphids. When aphids took up residence on my 'Autumn Joy' sedum, two tribes of ants – one red, one black – claimed them, and went to war on the plants. Ants kept grabbing other ants to hurl them off the plant.

The best way to deal with a situation like this – well, besides sitting down and watching the show – is to get the plant away from the ants, rob the aphids of protection, and allow natural predators to help you finish off the aphids.

I poured soapy water over the sedum, rubbing the leaves and stems to get all the aphids I could. A few drops of soap per gallon works just fine, and you can rinse the plant off with clear water afterward. Once the aphids were gone, the ant battles ceased – on that plant, anyway, since I didn't really keep track of the ants and their wars after that.

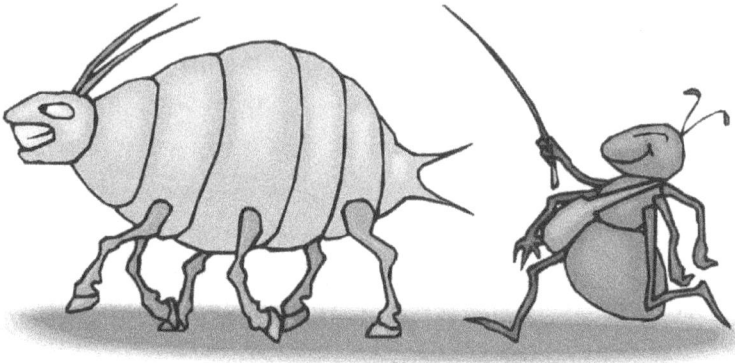

Once the ants are gone, you can discourage aphids. A good blast of water will knock them off the plant. Once the plant is dry, you can dust it with diatomaceous earth (also called DE). This is made up of microscopic diatoms, just like a piece of chalk, and the diatoms are sharp-edged, cutting through an insect's protective waxy layer (called chitin), which causes the insect to dry up and die. Diatoms are why a slug will explode if it crawls over a chalk line on the sidewalk. DE works well against any insect, good or bad, so be careful to keep this out of flowers when you're dusting the plant with it, so you don't accidentally kill off pollinators.

DE only works when it's absolutely dry, so apply it on dry days, and replace it after a rain. Don't breathe the dust, as it will also abrade your nasal linings and your lungs.

If you fertilize your plants regularly, cut back on the nitrogen. This element, which plants use to produce new growth, also makes the leaves soft and juicy for insect pests. Insects will jump on a succulent plant the way you'd jump on a juicy steak (or a sweet, juicy apple if you're a vegetarian).

A barrier of bone meal or charcoal will keep ants at bay – but only if the barrier stays dry. If your trees have ants running aphid farms on them, spray the tree with insecticidal soap, then put a barrier of Tanglefoot around the trunk of the tree. Tanglefoot is very sticky stuff and will stop the ants in their tracks. It will also stop any other insects, and possibly your cat, though I hope your cat will not stick to the tree like the ants will.

Spotted Cucumber Beetles

A spotted cucumber beetle chomping an innocent flower to shreds. Image by TimesArrow from Pixabay.

Coleoptera – Beetles

Spotted cucumber beetles, which look like green ladybugs, spread diseases to your cucumbers, melons, and cucurbits, and they will also spoil your rose blossoms.

These eat the leaves of cucumber plants and other cucurbits (like muskmelon and pumpkins). They can carry bacterial wilt from plant to plant while feeding, and this disease causes your plant to wilt, first in individual leaves, then nearby leaves, then through the whole plant. Sometimes the plant can succumb to the disease in only seven days.

If your plant is wilting and you think it might be bacterial wilt, try a simple test to be sure. Cut the stem and squeeze both cut parts. A sticky sap will ooze from it. If you stick the cut parts back together, then slowly draw them apart, you'll see ropes of sap stretch between them if the plant has bacterial wilt. Inside those ropes of sap are millions of bacteria. Eew.

Spotted cucumber beetles affect other plants. I usually find these beetles eating the stamens and petals of white or yellow roses. Then all the petals fall off prematurely. They seem to be

These beetles also spread squash mosaic virus and fusarium wilt. I wouldn't put it past them, to be honest.

Whenever I see one of these beetles, I just grab it in my fingers and squish it. If you don't want bug guts on your fingers, carry a spray bottle of insecticidal soap and spray the beetles when you see them. You might put floating row covers over your cucurbits (though watermelon is generally not affected) to keep the beetles off.

To keep numbers down next spring, till your garden in the fall. This may help expose beetle larvae and kill them. Clean up the garden and burn all cucurbit debris. Mulch your plants to deter the insects from laying eggs around your plants and slow larval migration through the soil. Another way to catch them is to put yellow plastic bowls filled with water and a little bit of dish soap near your plants. Cucumber beetles for some reason will crawl into the yellow bowls and drown in the dish soap water.

Tomatillos seem to be an effective trap plant for spotted cucumber beetle larvae. Let them grow in your garden until they start showing signs of larval damage – then pull them up and destroy them. The beauty part is that you can harvest fruit off the trap crops until you pull them up. They reseed so vigorously that you'll always have new plants popping up around the garden. If you are a tomatillo fan, then this solution may be just the thing for you!

Spider Mites

I hate electron microscope pictures of spider mites. This is a two-spotted spider mite feeding on a rose leaf, but ick ick ick. Photo by Eric Erbe, digital colorization by Chris Pooley, USDA Agricultural Research Service.

Spider mites are hard to spot, being only 1/50th of an inch long. Plants infested with spider mites generally turn a brownish-red, the leaves looking as if the sun burned them, or the leaves might look yellowish with green veins. When you turn the leaf over, you may see little webs and some tiny salt-and-pepper grit. On a plant heavily infested with spider mites, the leaves become bronzed and yellow and drop off.

Hot and dry weather can lead to a spider mite infestation. Spider mites infect some roses (I had an antique rose called 'Rose de Rescht' that kept catching them), as well as on burning bushes, tomatoes, and marigolds. In advanced cases, the plant becomes swathed in webs until it turns into a plant mummy.

To see if your plant has mites, shake the leaves over white paper – the tiny mites will fall on the paper and creep around. If you crush them with your finger, the plant-feeding mites will leave green streaks, while the beneficial predator mites will leave yellow or orange streaks. You want to keep the beneficial mites, naturally. If you get ten or more spider mites on the paper every time you shake the foliage, you should probably take steps to control them.

Control spider mites with a strong spray from a garden hose from underneath. This disrupts their activity, washes off the dust that hides the mites from natural predators, and helps to keep the spider mites under control in the landscape. The jets of water will knock some leaves off – but these leaves were going to fall off anyway.

Another way is to spray horticultural oil, or dormant oil, on the affected plant. This suffocates the mites with a thin layer of oil that still allows the plants to breathe. It's also a safe way to kill the mites, as opposed to chemicals. If you spray horticultural or dormant oil, do it when it's cool, to try and keep the leaves from burning.

Some plants blotch when they have this oil sprayed on them; check the label before you buy it. Avoid spraying flowers.

Prune off badly infested branches and leaves and dispose of them, though not in the compost, because the heat of the compost will not kill these mites.

A kaolin clay spray would suffocate the spider mites – give it a try.

Thrips

Thrips on a flower petal. Photo by Ryan Hodnett.

Thrips are tiny insects that are small enough to be drifted by wind from plant to plant, use a tiny proboscis (i.e. a bug straw) to puncture the leaf and suck out the contents of the cells, causing the surface of the leaf to turn silvery or whitish. When thrips feed on rose petals, you see unsightly brown spots. Unfortunately, by the time you notice these spots, the thrips are usually long gone. They do their damage while the rose blossom is still in bud, so you don't see their work until the flower opens. By then, they've already laid their eggs and died. However, thrips can have as many as eight generations per year, depending on conditions.

Here's an interesting bit of news via *Fine Gardening* magazine. One interesting way to control thrips is to allow aphids to show up on your roses.

According to an organic rose nursery, Bierkreek, the hoverfly (a small fly that disguises itself as a bee, and is often mistaken for a sweat bee) is very beneficial to roses. In spring, the hoverfly lays its eggs near aphids. When the hoverfly larvae hatch out, they eat the aphids. Once these metamorphose into adults, they'll lay their eggs – which hatch when thrips are out and about. And the second generation of hoverfly larvae eat thrips!

How do you know that this hoverfly is a fly instead of a bee? First, flies have big, googly eyes. Bees and wasps have small eyes. Second, nearly all flies have only one set of wings. Bees have two. Third, bees and other members of their family have the thin "wasp" waist. Flies don't have this waist.

Encouraging predatory insects are the best way to control thrips in the first place. On hot days, spray your roses with water to wash off the dust that shelters thrips. If you've had trouble with thrips already, you might spray a little bit of Neem oil onto your buds before they open – keep it off already opened flowers, so you don't get any on any visiting honeybees. Spray the buds every other day due to the thrips' fast life cycle. Cut off and dispose of any thrip-damaged blossoms to get rid of any eggs or thrips that might be on them.

Japanese Beetles

A Japanese beetle. Note the hind legs in the air. It does this when it's preparing to fall off the plant. Image by Richard Malo from Pixabay.

When I was taking care of the rose garden in Missouri, I didn't have any trouble with Japanese beetles. Then I started going to St. Paul, Minnesota every summer, and I was agog at the mobs of Japanese beetles on their rose blossoms. I'd grab the beetles off the rugosas and crush them in my fingers, but just as soon as I turned around there'd be more and more landing on the roses. Ugh! They were like tiny nanobot creatures intent on the destruction of all the world's roses! And worse, they were busy skeletonizing the leaves of various other plants.

And now, this year after we had a cold, wet spring, they started showing up in huge numbers on my backyard Missouri roses in June, finally dwindling away in early August. Now they seem to be here to stay.

So what do you do when the nanobot apocalypse looms for your roses?

One way to stop Japanese beetles is by handpicking them. In the early morning, when beetles are sluggish, knock them off your plants into a bucket of soapy water. (You can catch them through the rest of the day in this way, but you have to be quicker, because they fly away as you're reaching for them.) You can also spray them with Neem oil products. (Be sure these contain azadirachtin, also called BioNeem; also, look for the active ingredient "clarified hydrophobic extracts of neem.") Insecticidal soap is generally ineffective.

It's important to keep handpicking, because when adult Japanese beetles congregate on plants, they attract more and more beetles.

If you have only a few roses, you might wrap your blossoms in cheesecloth to keep the little jerks away from them.

Pheromone traps are a possibility – except that they end up bringing in more Japanese beetles than they kill. These work best if you're able to put them in a place far away from your plants. Empty out the bag daily, because dead beetles really stink.

Chickens love to eat Japanese beetles, and you can find various ideas online about how to rig a pheromone trap to feed your hungry flock. What I've done is put up a pheromone trap in the chicken pen over a big pan of water, and I cut the bottom out of the bag under the trap. The clumsy beetles fall through the bag into the water and struggle until a chicken very happily puts them out of their misery. Some chickens will even jump and catch beetles that are on their way to the pheromone traps. The chickens will snap up the beetles in a very satisfying way, and when I have the cup out, traveling from tree to tree in my yard, Henny Penny will follow me, gazing up at me very hopefully every time I'm knocking beetles in the cup.

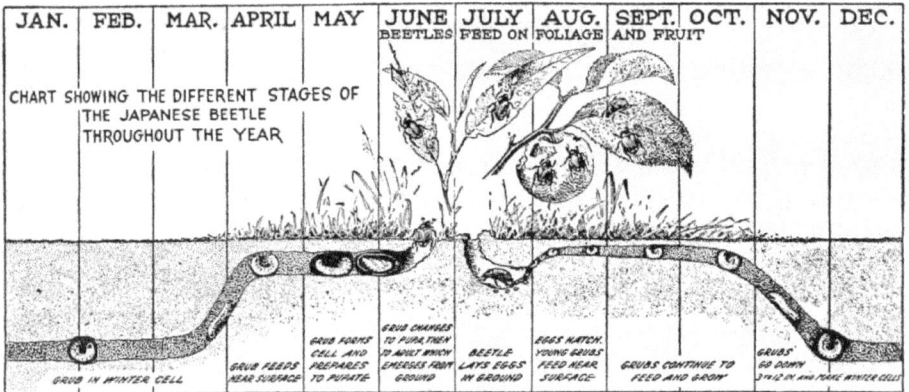

JAN.	FEB.	MAR.	APRIL	MAY	JUNE	JULY	AUG.	SEPT.	OCT.	NOV.	DEC.
					BEETLES	FEED ON	FOLIAGE	AND FRUIT			

CHART SHOWING THE DIFFERENT STAGES OF THE JAPANESE BEETLE THROUGHOUT THE YEAR

An illustration that shows the life cycle of the Japanese beetle.

A nice way to kill two birds (beetles?) with one stone is to wear Spikes of Death (also called Aerator Sandals) around the yard. These are basically sandals with big spikes in the bottoms. You have to walk very carefully in them so you don't accidentally bust an ankle! If you walk all over your yard, the spikes puncture the grubs that eventually will turn into Japanese beetles. Use the spikes in April and May. If your lawn has grub problems, you might be able to put a small dent in the local Japanese beetle population, kill off many of the grubs and pupating beetles in your lawn, and aerate your soil.

Note: If you're being afflicted by Japanese beetles, I've written a whole book about different ways to eradicate them called Japanese Beetles and Grubs: Trap, Spray, and Control Them, which can be found on most online book retailers.

Squash Bugs

Squash bugs look like alien invaders from some Hollywood B-film. The adult bugs are grey or brown, and their backs look like a shield. They are often misidentified as stink bugs, but these are longer than stink bugs, which are shield-shaped and squat. Fortunately, they're both in the same order (Hemiptera), so their control methods will be similar.

Squash bugs can be found on your squash, watermelon, and cucumber vines in summer. A large number of squash bugs feeding on the plants can cause the vines to wilt badly, but the wilt ends when you get rid of the bugs.

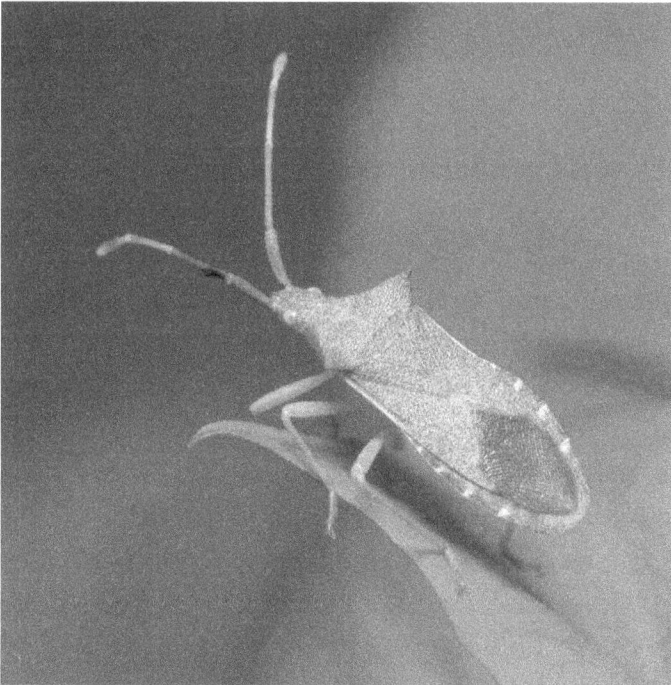

To nip the squash bugs in the bud, lift up the squash leaves and crush the copper-colored egg clusters underneath. Clean up plant debris so the bugs have no place to hide. Lay several boards around the plants. Every morning, lift the boards and kill any squash bugs hiding underneath.

A raft of squash bug eggs.

If the bugs are getting out of control, spray insecticidal soap or pyrethrum (an organic insecticide made from chrysanthemums), but this works best only when the bugs are still small. You have to spray pyrethrum directly on the pests, but they fall off the plant and die as soon as it hits them. Sometimes I carry a bottle with me so I can zap those green ladybugs in the rose blossoms, or the tobacco hornworms on the Nicotine flowers.

At the end of the season, destroy all vines and clean up the garden so adults have no place to spend the winter.

Next year, cover vines with a fine-mesh cloth, such as floating row covers, to keep bugs out. Once flowering begins, remove the cloth so bees can pollinate the flowers. By then it will be too late for squash bugs to do their damage.

Grasshoppers

When the plagues of grasshoppers start, it seems that nothing will stop them outside of dynamite. However, when you start working on killing off the grasshoppers in May, when they're small, you have a greater chance of slowing down the onslaught and protecting your vegetables and plants.

Get 'em when they're small

Grasshoppers are easiest to kill when they're less than ½ inch long. Start early and start looking for grasshopper hatching grounds. Good places to look for hatching grounds are ditches, fence rows, pastures, or roadsides. If there is a large number of small grasshoppers in one area, you're in the right place. These young grasshoppers are called nymphs, and look just like full-sized grasshoppers except they are ½ to ¾ inch long. When in the nymph stage, grasshoppers don't move very far from where they've hatched. If you find hatching grounds in your area, flag them, spray them, and check back on them later.

You can also use *Nosema locustae* bait at the hatching grounds. *Nosema locustae* is a protozoan microbe that causes diseases in grasshoppers. The protozoans are mixed in a bait that the grasshoppers eat – it's available as Nolo Bait and as Semaspore. Infected grasshoppers are cannibalized by non-infected grasshoppers, which in turn get cannibalized in one of those delightful "Nature Gets Gross" turnabouts.

If you use the bait correctly, while the hoppers are still small, you can get a 30 to 40 percent kill rate. However, Nolo Bait isn't as effective on mature grasshoppers. Spread the bait when the weather is mild for best results. But when you set out the bait in the hatching grounds, it does work. Also, broadcast some bran bait between the hatching grounds and your garden to catch the young hoppers moving in. Nolo Bait works best if you use it correctly for several years in a row. The disease will carry over into the next year.

Protip: Every insect pest, weed, or disease is most effectively dealt with when they're small, or just getting established.

A common grasshopper, Phoetaliotes nebrascensis. Photo by Stephen Ausmus, USDA Agricultural Research Service.

How to control grasshoppers when they're big

Once the grasshoppers get big enough to start flying, usually about late June, then you're in more trouble, because now they get hard to eradicate. Insecticides and sprays still work; however, when you kill the grasshoppers in the garden, more hoppers will migrate in to take their place. It seems you could spray all summer and not get anywhere.

To keep the grasshoppers from entering your yard, use a trap crop to catch the grasshoppers as they travel in from surrounding pastureland. Grow a tall, lush stand of grass around your yard. Fertilize it with nitrogen to make the grass really succulent and tasty, and water it. The grasshoppers will latch on to the grass and start eating that. Then you can spray the grasshoppers inside the trap crop as they eat.

One note: don't let the trap crop of grass dry out, and don't mow it, because then the grasshoppers will head to the next oasis of green: your garden.

Other means of grasshopper destruction

Organic insecticides such as insecticidal soap or neem oil will work against the grasshoppers, but only if you hit the actual insect with the spray. Floating row covers also protect your garden against grasshopper onslaughts. However, if you have crops such as cucumbers or squash under them, the plants will have to be hand-pollinated in order to bear fruit, since the cloth also keeps bees out. Also, if the grasshopper infestation is really bad, they may eat their way through the cloth! In that case, you may have to use metal window screening to keep them out. Or maybe dynamite.

Desperate times call for desperate measures

If you live in the country, get some poultry. Guinea hens are best, but if you prefer peace and quiet to the machine-gun rattle of a bunch of guinea hens shouting *"Po-quak, po-quak, po-quak"* at the tops of their lungs, you'd better settle for some chickens or ducks.

Some birds also feed on grasshoppers, including bluebirds, mockingbirds, brown thrashers, crows, and sparrows. Attract birds to the yard with a clean birdbath, good habitat, and some hiding places.

Hand-picking the hoppers can help, especially if you need a lot of good fish bait.

The situation gets more serious if the grasshoppers are moving in from surrounding pastures that have had no treatment, especially once the pastures start browning in the summer heat. And most treatment beyond hand-picking is ineffective against flying adults. Sometimes I cut them in half with my garden shears. Once, the back half of a hopper jumped about five feet into the air after I cut it in two. It would have been really interesting if I hadn't been so busy gagging at the sight.

Emerald Ash Borer

Coleoptera – Beetle family

I apologize to my paperback readers for the black and white photos (color printing makes this book cost more), but this emerald ash borer should be as green as the Emerald City in The Wizard of Oz. Photo courtesy of the USDA on Flickr.

When I was taking my usual walk during break earlier this year, I was struck by a green ash tree that was along the route. "Wait a minute," I thought, squinting at the top of the tree. "That's dieback." The ash tree, otherwise green and leafy and healthy, was suddenly losing leaves at the ends of its branches, out of nowhere. *Is this what emerald ash borers do to an ash tree?* I wondered.

As it turned out, this was. I had not seen a tree with borers before, but it didn't take long for me to see what they could do. To be honest, I was stunned at how fast the tree went downhill, losing all its leaves within the space of a month or two. Then other ash trees in the neighborhood swiftly died from the same insects. I informed the homeowners when I ran into them outside and told them what was happening to their trees, but at this point there wasn't much they could do to save them, which was sad because these were lovely trees.

The emerald ash borer (*Agrilus planipennis*), which is a brilliant green beetle no bigger than a dime, is a relatively new beetle that first showed up on our shores in 2002, and it started spreading as soon as it appeared, killing off ash trees. The adult beetles don't cause much damage – these thin beetles will eat a few leaves, but mostly they're interested in mating.

The real damage is done by the beetle larvae, which feed on the inner bark of the ash trees, the cambium layer, which transports water and nutrients from the leaves and roots to the rest of the tree. The larvae will create long, serpentine tunnels through the living layers of the tree. As the cambium layer is eaten away, girdling the tree, the leaves at the extremities of the tree start dying off, and the damage quickly moves through the rest of the tree.

If you see your ash tree dying back, look at the bark. There will be small holes that look like D-shaped drill holes in the bark – this is where the newly metamorphosed adult chews its way out of the tree after spending all winter inside the tree as a pupa.

In advanced cases, you'll see that the bark has been "blonded" – the gray top layer of the outer bark has been chewed off, leaving it a blonde color. If you cut off the outer bark and peel it back, you'll see little tunnels in curvy designs up and down the inner bark of the tree.

If you ever buy firewood, here's one way to keep the emerald ash borer from spreading: Buy all your firewood locally. Beetle infestations have been spread across state lines by people transporting firewood with emerald ash borers hidden in the bark of the logs. The same goes with nursery stock or the trees you're buying.

Even if you have healthy ash trees, they can still succumb and die from this borer. A few trees have shown some resistance to this insect, and breeders are already working to develop ash species that can withstand borer attacks. But this will take time, unfortunately.

So! That's it for the bug book! I hope this book has been helpful to you. If it has, do leave a review and let other readers know. Reviews are gold to authors.

OTHER BOOKS IN THE HUNGRY GARDEN SERIES

Little Pots, Big Yields: Container Gardening for Creative Gardeners

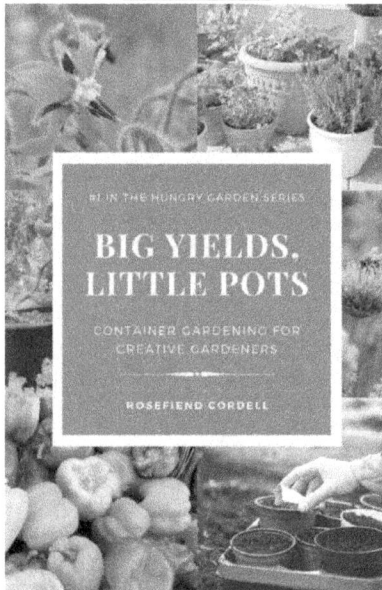

If you are short on space, or if you have the soil from hell, or if you have a hard time stooping and bending, then growing vegetables in containers is the solution for you. If you want to talk about improving the quality of your life, the fresh herbs and tomatoes and strawberries ripening on your balcony will do the job.

Your container vegetable garden will take a small investment of time and effort, but anything good does. Patience and practice in gardening will yield the best results.

This book covers: Choosing the right container * How to start seeds (and combat damping-off disease) * Soilless mixes and their elements * Fertilizer, watering, climate, trellising

And this book will dig into the different kinds of vegetables that grow best in pots - best methods for each crop - best varieties for containers. This book is the essentials guide to container gardening for beginners and also for seasoned gardeners who have been around the block a few times.

Edible Landscaping: Foodscaping and Permaculture for Urban Gardeners

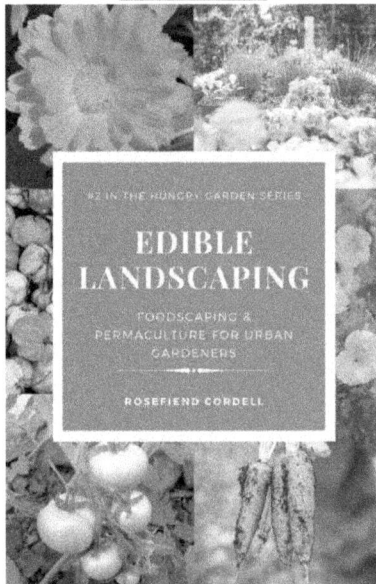

Liberate your food plants from the vegetable garden! Landscape your living space so it offers food for the eyes, heart, and stomach. Edible Landscaping: Foodscaping and Permaculture for Urban Gardeners is a how-to gardening manual written by a hard-boiled former horticulturist who hates weeding with the heat of a million suns. Rosefiend Cordell takes the budding gardener on a step-by-step process to transform their sad yard into a merry garden full of ornamental flowers linking arms with tomatoes, herbs, and edible flowers, as well as good fruit and nut trees.

This gardening book features practical gardening methods that help you create a design to build the outdoor living space you want. Information on foodscaping and permaculture, and how these techniques can help you to build the soil, prepare a garden design, and choose the plants you want.

Create a mixed border that cuddles herbs, edible flowers, vegetables, ornamental plants, and fruits together in harmony. It doesn't matter if you have a brown thumb or a green thumb. If you live in the city, the suburbs, or way out in the sticks, this handy-dandy manual will teach you how to make the best use of the space you have while opening your eyes to a great old way of gardening that's beautiful, tasty, and deeply satisfying.

Now available at your favorite online bookseller!

Indoor Gardening: Growing Herbs, Greens, and Vegetables Under Lights

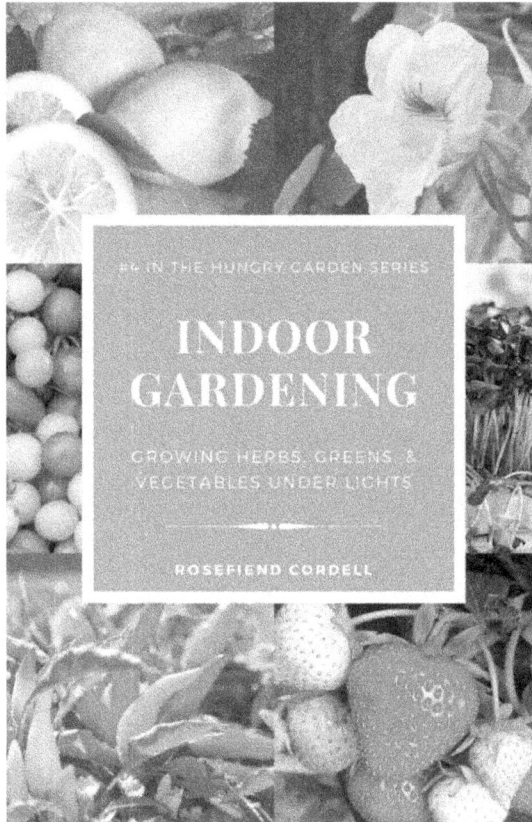

Preorder is available!!

Growing a Food Forest: Trees, Shrubs, and Perennials That'll Feed Ya!

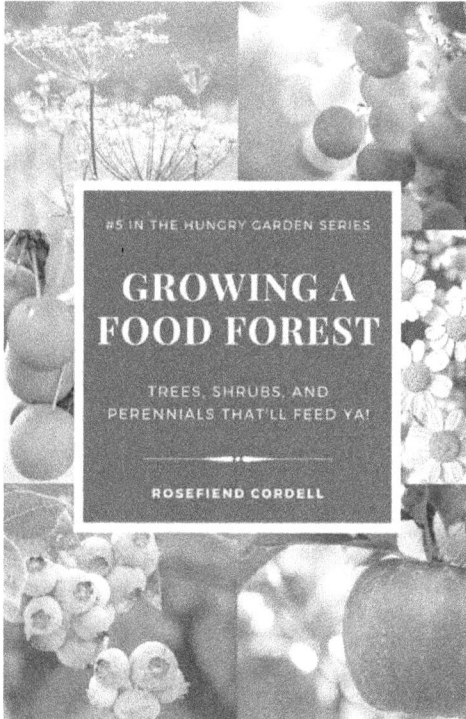

Preorder your copy here!
Out on 1 September 2023.
https://books2read.com/u/mV0wzp

A PREVIEW OF THE EASY-GROWING GARDENING SERIES

The Easy-Growing Gardening series is 13 books written to help you navigate the garden. Need help with the rose garden? Growing perennials? Garden design? Vegetable gardening? Tomatoes? Are Japanese beetles getting you down? These books will help you with these topics and more.

Don't Throw in the Trowel: Vegetable Gardening Month by Month.

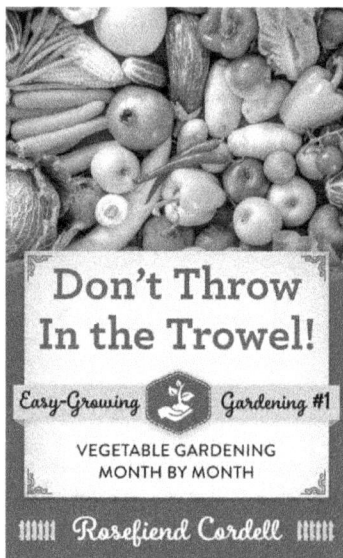

Two things:
First: *You know more about gardening than you think.*
Second: *A garden – the soil – plants – all of these are very forgiving. When it comes down to it, you can make a lot of mistakes and still come out with good results.*

Don't Throw in the Trowel! Vegetable Gardening Month by Month includes info on seeds, transplanting, growing, and harvesting, as well as diseases, garden pests, and organic gardening. I also talk about garden prep, because a good plan, a garden notebook, and a little off-season work will save you a lot of trouble down the road.

I've worked in horticulture for 20 years: in landscape design and installation, as a greenhouse tech, perennials manager, and city horticulturist & rosarian. This book shares what I've learned so far.

Keep reading for a little sample from the first book in the series ... *Don't Throw in the Trowel: Vegetable Gardening Month by Month*.

Save Time and Trouble with Garden Journals

When I worked as a municipal horticulturist, I took care of twelve high-maintenance gardens, and a number of smaller ones, over I-don't-know-how-many square miles of city, plus several hundred small trees, an insane number of shrubs, a greenhouse, and whatever else the bosses threw at me. I had to find a way to stay organized besides waking up at 3 a.m. in a panic to make extensive lists. My solution: keep a garden journal.

Vegetable gardeners with an organized journal can take control of production and yields. Whether you have a large garden or a small organic farm, it certainly helps to keep track of everything in order to beat the pests, make the most of your harvest, and keep up with spraying and fertilizing.

Keeping a garden journal reduces stress because your overtaxed brain won't have to carry around all those lists. It saves time by keeping you focused. Writing sharpens the mind, helps it to retain more information, and opens your eyes to the world around you.

My journal is a small five-section notebook, college ruled, and I leave it open to the page I'm working on at the time. The only drawback with a spiral notebook is that after a season or two I have to thumb through a lot of pages to find an earlier comment. A small three-ring binder with five separators would do the trick, too. If you wish, you can take out pages at the end of each season and file them in a master notebook.

I keep two notebooks – one for ornamentals and one for vegetables. However, you might prefer to pile everything into one notebook. Do what feels comfortable to you.

These are the five sections I divide my notebooks into – though you might use different classifications, or put them in different orders. Don't sweat it; this ain't brain surgery. Feel free to experiment. You'll eventually settle into the form that suits you best.

First section: To-do lists.

This is pretty self-explanatory: you write a list, you cross off almost everything on it, you make a new list.

When I worked as horticulturist, I did these lists every month. I'd visit all the gardens I took care of. After looking at anything left unfinished on the previous month's list, and looking at the garden to see what else needed to be done, I made a new, comprehensive list.

Use one page of the to-do section for reminders of things you need to do next season. If it's summer, and you think of some chores you'll need to do this fall, make a FALL page and write them down. Doing this has saved me lots of headaches.

Second section: Reference lists.

111

These are lists that you'll refer back to on occasion.

For example, I'd keep a list of all the yews in the parks system that needed trimmed, a list of all gardens that needed weekly waterings, a list of all places that needed sprayed for bagworms, a list of all the roses that needed to be babied, etc.

I would also keep my running lists in this section, too – lists I keep adding to.

For instance, I kept a list of when different vegetables were ready for harvest – even vegetables I didn't grow, as my friends and relatives reported to me. Then when I made a plan for my veggie garden, I would look at the list to get an idea of when these plants finished up, and then I could figure out when I could take them out and put in a new crop. I also had a list of "seed-to-harvest" times, so I could give each crop enough time to make the harvest date before frost.

You can also keep a wish list – plants and vegetables you'd like to have in your garden.

Third section: Tracking your progress.

This is a weekly (or, "whenever it occurs to me to write about it") section as well.

If you plant seeds in a greenhouse, keep track of what seeds you order, when you plant them, when they germinate, how many plants you transplant (and how many survive to maturity), and so forth.

When you finish up in the greenhouse, use these pages to look back and record your thoughts – "I will never again try to start vinca from seeds! Never!! Never!!!" Then you don't annoy yourself by forgetting and buying vinca seeds next year.

You can do the same thing when you move on to the vegetable garden – what dates you tilled the ground, planted the seeds, when they germinated, and so forth. Make notes on yields and how everything tasted. "The yellow crooknecks were definitely not what I'd hoped for. Try yellow zucchini next year."

Be sure to write a vegetable garden overview at season's end, too. "Next year, for goodness' sake, get some 8-foot poles for the beans! Also, drive the poles deeper into the ground so they don't fall over during thunderstorms."

During the winter, you can look back on this section and see ways you can improve your yields and harvest ("The dehydrator worked great on the apples!"), and you can see which of your experiments worked.

Fourth section: Details of the natural world.

When keeping a journal, don't limit yourself to what's going on in your garden. Track events in the natural world, too. Write down when the poplars start shedding cotton or when the Queen's Anne Lace blooms.

You've heard old gardening maxims such as "plant corn when oak leaves are the size of a squirrel's ear," or "prune roses when the forsythia blooms." If the spring has been especially cold and everything's behind, you can rely on nature's cues instead of a calendar when planting or preventing disease outbreaks.

Also, by setting down specific events, you can look at the journal later and say, "Oh, I can expect little caterpillars to attack the indigo plant when the Johnson's Blue geranium is blooming." Then next year, when you notice the buds on your geraniums, you can seek out the caterpillar eggs and squish them before they hatch. An ounce of prevention, see?

113

When I read back over this section of the journal, patterns start to emerge. I noticed that Stargazer lilies bloom just as the major heat begins. This is no mere coincidence: It's happened for the last three years! So now when I see the large buds, I give the air conditioner a quick checkup.

Fifth section: Notes and comments.

This is more like the journal that most people think of as being a journal – here, you just talk about the garden, mull over how things are looking, or grouse about those supposedly blight-resistant tomatoes that decided to be contrary and keel over from blight.

I generally put a date on each entry, then ramble on about any old thing. You can write a description of the garden at sunset, sketch your peppers, or keep track of the habits of bugs you see crawling around in the plants. This ain't art, this is just fun stuff (which, in the end, yields great dividends).

Maybe you've been to a garden talk on the habits of Asian melons and you need a place to put your notes. Put them here!

This is a good place to put garden plans, too. Years later I run into them again, see old mistakes I've made, and remember neat ideas I haven't tried yet.

Get a calendar!

Then, when December comes, get next year's calendar and the gardening journal and sit down at the kitchen table. Using last year's notes, mark on the calendar events to watch out for -- when the tomatoes first ripen, when the summer heat starts to break, and when you expect certain insects to attack. In the upcoming year, you just look at the calendar and say, "Well, the squash bugs will be hatching soon," so you put on your garden gloves and start smashing the little rafts of red eggs on the plants.

A garden journal can be a fount of information, a source of memories, and most of all, a way to keep organized. Who thought a little spiral notebook could do so much?

Would you like to read more? <u>Grab a copy here</u>**!**

USDA PLANT HARDINESS ZONE MAP

Now and then I mention what zone a plant is hardy to. For those of you who reside in the United States, the hardiness zone is depicted on this map, and shows the high and low temperatures of the area where you live. This map is from 2012, and as you know, climate change is still throwing this thing off to some extent, and will be changing it more over in upcoming years. But it's still helpful.

The USDA has this map on their website, but if you go there, you can click on your state to get a closer view of what your particular zone looks like. Go to their website over here at https://planthardiness.ars.usda.gov/PHZMWeb/ to get started.

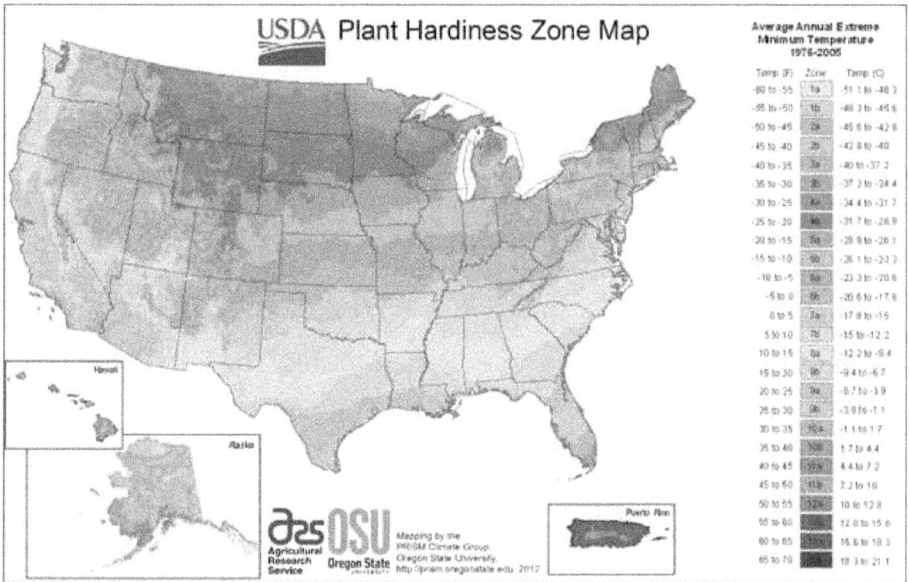

USDA Plant Hardiness Zone Map

Average Annual Extreme Minimum Temperature 1976-2005

Temp (F)	Zone	Temp (C)
-60 to -55	1a	-51.1 to -48.3
-55 to -50	1b	-48.3 to -45.6
-50 to -45	2a	-45.6 to -42.8
-45 to -40	2b	-42.8 to -40
-40 to -35	3a	-40 to -37.2
-35 to -30	3b	-37.2 to -34.4
-30 to -25	4a	-34.4 to -31.7
-25 to -20	4b	-31.7 to -28.9
-20 to -15	5a	-28.9 to -26.1
-15 to -10	5b	-26.1 to -23.3
-10 to -5	6a	-23.3 to -20.6
-5 to 0	6b	-20.6 to -17.8
0 to 5	7a	-17.8 to -15
5 to 10	7b	-15 to -12.2
10 to 15	8a	-12.2 to -9.4
15 to 20	8b	-9.4 to -6.7
20 to 25	9a	-6.7 to -3.9
25 to 30	9b	-3.9 to -1.1
30 to 35	10a	-1.1 to 1.7
35 to 40	10b	1.7 to 4.4
40 to 45	11a	4.4 to 7.2
45 to 50	11b	7.2 to 10
50 to 55	12a	10 to 12.8
55 to 60	12b	12.8 to 15.6
60 to 65	13a	15.6 to 18.3
65 to 70	13b	18.3 to 21.1

Hawaii

Alaska

Puerto Rico

Agricultural Research Service

Oregon State

Mapping by the PRISM Climate Group, Oregon State University, http://prism.oregonstate.edu, 2012

117

Me in February 2018 with two baby chickies, my laptop, and a can of Red Bull.
This is how I roll, people.

ABOUT THE AUTHOR

A former city horticulturist and a long-time garden writer, Rosefiend Cordell, aka Melinda R. Cordell, has written 12 books in the Easy-Growing Gardening series under the name Rosefiend Cordell, and three books (so far) in the Hungry Garden series.

She's worked in horticulture for half of her life – longer if you count when she was young, collecting wildflowers. She's worked in greenhouses, both retail and commercial; as a landscape laborer and designer, as a perennials manager, as municipal horticulturist and public rose garden potentate, and now as a gardening author (which is much easier on the back and joints).

Melinda R. Cordell has written a truckload of YA novels, including the Dragonriders of Fiorenza series. Set in an alternative medieval Italy, it features a wily dragonrider, her loyal dragon, and her assassin grandma, all pitted against a world out to strip away every last one of their hopes and dreams.

Melinda lives in northwest Missouri with her husband and two kids, the best little family to walk the earth, and is writing about 24 books at once, fueled by passion and caffeine.

If you want to keep up with her, you can drop her a friendly note at rosefiend@gmail.com.

Don't forget to leave a book review on your favorite retailer, BookBub, or Goodreads!

melindacordell.com

www.ingramcontent.com/pod-product-compliance
Lightning Source LLC
Chambersburg PA
CBHW030842090426
42737CB00009B/1076